T0290942

IET ENERGY ENGINEERING SERIES 118

Modeling and Simulation of Complex Power Systems

Other volumes in this series:

Modeling and Simulation of Complex Power Systems

Antonello Monti and Andrea Benigni

The Institution of Engineering and Technology

Published by The Institution of Engineering and Technology, London, United Kingdom

The Institution of Engineering and Technology is registered as a Charity in England & Wales (no. 211014) and Scotland (no. SC038698).

The Institution of Engineering and Technology
Futures Place
Kings Way, Stevenage
Hertfordshire, SG1 2UA, United Kingdom

www.theiet.org

British Library Cataloguing in Publication Data
A catalogue record for this product is available from the British Library

ISBN 978-1-78561-404-0 (hardback)
ISBN 978-1-78561-405-7 (PDF)

Typeset in India by MPS
Printed in the UK by CPI Group (UK) Ltd, Croydon

Contents

About the authors

Antonello Monti is the professor–director at the Institute for Automation of Complex Power Systems, RWTH Aachen University, Germany. Prior to this, he worked at the University of South Carolina (USA), where he was associate director of the Virtual Test Bed (VTB) project on computational simulation and visualisation of modern power distribution. Since 2019, he has also a joined appointment at Fraunhofer FIT as part of the Center for Digital Energy Aachen.

Andrea Benigni is a full professor at RWTH-Aachen and director of the Institute of Energy and Climate Research: Energy Systems Engineering (IEK-10) at the Juelich Research Center, Germany. He received the B.Sc. and M.Sc. degrees from Politecnico di Milano, Milano and the Ph.D. degree from RWTH-Aachen University, Aachen, Germany. From 2014 to 2019, he was an Assistant Professor with the Department of Electrical Engineering, University of South Carolina, Columbia, SC, USA.

Supplementary material

Supplementary material for this book can be obtained by:

- downloading from the IET Digital Library at https://digital-library.theiet.org/content/books/ra/pbpo118e. Once on the page, click on the "Supplementary Material" tab to access the files.
- alternatively email books@theiet.org to be sent a copy of the files.

Additional contributors

The below authors made a valuable contribution to the book.
Chapter 6: Matthew Milton, University of South Carolina
Chapter 7: Markus Mirz and Jan Dinkelbach, RWTH Aachen University
Chapter 8: Markus Mirz and Jan Dinkelbach, RWTH Aachen University
Chapter 9: Prof. Ferdinanda Ponci, RWTH Aachen University
Chapter 10: Christian Dufour, Ph.D. and Senior Simulation Specialist, Power Systems and Motor Drives, OPAL-RT Technologies and Jean Belanger, Founder, CTO and CEO of OPAL-RT Technologies

Chapter 1

Introduction

Antonello Monti[1,2] and Andrea Benigni[3,4]

Simulation is a key activity in every engineering domain. It is basically impossible today to imagine a design activity that does not include a simulation verification.

The role of simulation has been actually growing more and more in practice. While the main role of simulation in the past has been to replicate reality, it has been more and more moving to anticipate reality. While in the past, it was mostly used to better understand the existing systems, it is now driving the design process so that we may say that the target is now on reality to replicate simulation. Today every complex modern system is first built-in simulation and then realized in reality. Simulation becomes then a formal method of specification that can better summarize the requirements, thanks to the fact that it is strictly a mathematical process.

Nevertheless, simulation is also still used to better understand the systems that we already built and operate. A simulation model can be for example used to define and analyze "what if" scenarios for an infrastructure.

Simulation in the power system has always been a very important activity, mostly because an experimental activity, strictly speaking, is basically impossible. While testing can be performed on a real grid, it is anyway impossible to perform comprehensive testing or testing that can impact wide areas such as a full transmission system. Also, simulation has been always offering a safer way to understand critical situations such as faults.

Real-time simulation has in the recent years added a new dimension to the power system research offering close to reality experiments in a different scale from small-field tests. The further development in the direction of hardware in the loop and power hardware in the loop has extended the concept of testing and validation in the power engineering domain.

All these considerations should make clear how the simulation science is a key asset for a modern power engineer. On the other hand, when we deal with

[1]Institute for Automation of Complex Power Systems, RWTH Aachen University, Germany
[2]Fraunhofer FIT Center for Digital Energy, Germany
[3]Institute of Energy and Climate Research: Energy Systems Engineering (IEK-10), Juelich Research Center, Germany
[4]Department of Mechanical Engineering, RWTH Aachen University, Germany

simulators, we typically deal with a sort of black box. As good engineering practice it is then critical to have an understanding of what is inside the box.

In the trend to simulation-driven engineering, it is in fact very important to deeply understand how simulators work to be sure that we always take educated decisions in the process. Building trust in the simulation results is a key activity for every engineer. A modern engineer is not likely to develop a new simulation platform but it is definitely likely to use one or more of them for the everyday job.

The purpose of this book is to look "under the hood" of modern simulators to help engineers developing their understanding of when and why the simulations results from a given tool can be trusted.

In effect, there is not a single simulator able to tackle all the questions that a power engineer may face, but there are several tools that are better suited for a given question.

Different commercial platforms use different modeling approaches and each approach may face limitations in a given condition. In this respect, this book will not define the perfect solution for every question, but it will provide an unbiased guide to different simulation approaches presenting pros and cons of the different solutions.

1.1   The structure of the book

As already introduced, the book focuses on simulation methods trying to give a complete coverage of the most used methodologies.

Chapter 2 is an introduction to digital simulation. Here we do not enter in the methodologies, while we focus on issues related to the digital execution of the models. A review of the typical discretization approaches offers the reader the possibility to appreciate challenges in terms of complexity versus accuracy.

In the following chapter, the topological methods are introduced. Among the topological methods, we decided to focus on nodal analysis. Nodal analysis is the base of some of the most famous circuit simulators such as SPICE. The automatic solution of circuit based on nodal analysis and future development is a cornerstone of the simulation science. It was mostly developed during the 1970s when the basic idea of resistive companion was introduced both in the microelectronic arena and in the power system domain. The chapter starts with a consideration on linear circuits, but also non-linear methods are introduced at the end.

Chapter 4 focuses on state space methods. Somehow state space has been always a competitor of the topological methods. State space comes more from the control community but has been successfully applied also in circuit analysis. Even here, several commercial examples are available. It is then a reasonable question to ask if there is a way to combine the two approaches. This has been done in several ways and an overview of the options completes the review of the state equation approach.

No matter what method we adopt sooner or later, we reach a limit on the computational capability of a given platform and the question of parallelization arises. Chapter 5 presents some more classical and some more recent options to parallelize the execution of a power system simulator.

The application of simulation to the design process has also increased the interest in incorporating uncertainty in the modeling process. Uncertainty in the design may be intentionally introduced to model and analyze tolerance but it can also be the result of specifications that will be detailed at a later stage. Chapter 6 reviews some of the methods that support the introduction of uncertainty starting from the classical Monte Carlo approach and coming to more advanced tools such as polynomial chaos.

In following chapters, the book explores other peculiar cases of simulation. First of all, the focus is on simulation languages. Many times we confuse the concept of modeling with the simulation itself. The creation of modeling languages has made this distinction definitely clear. Modelica is probably the most significant example in this direction. While it was developed originally for mechanical engineering applications, it is now earning a significant role also in electrical engineering and power engineering in particular.

While the first part of the book was fundamentally focused on the so-called electromagnetic type of simulation, power system engineers also need other types of analysis that allows the execution of models with a larger time step, so that we can focus on longer transient. A classical approach to electromechanical simulation is introduced but the chapter presents also more recent developments that are trying to bridge between the mechanical and the electromagnetic domain. The pervasive presence of power electronics is making the grids more dynamic and the borders of these two analysis are not so clear anymore. Dynamic phasors are emerging as an interesting alternative that fits perfectly in the middle.

On the other hand of the time scale is vice versa, the detailed analysis of power electronic circuits. Here we face highly non-linear and time-variant circuits. How to address the complexity of power electronics so that it can be incorporated in power system analysis is the topic of the Chapter 9.

In conclusion, real-time simulation and hardware in the loop are introduced to cover the most advanced role that the simulation is playing: also from the time perspective, substitute the reality with a model that is behaving coherently.

1.2 How to use the book

The structure of the book is inspired by the content of the class "Modeling and Simulation of Complex Power Systems" as offered since now 10 years at RWTH Aachen University at Master Level. The content of the book can also serve the teaching at bachelor level skipping some of the most advanced concepts such as simulation under uncertainty.

The book can also serve as an everyday reference for professionals dealing with simulation challenges to offer a quick overview of the methods, their limits, and their advantages.

The text is also supported by SW implementation of most of the algorithms to give the reader the chance to see down to the code level how the simulation can be performed.

Supplementary material

Supplementary material for this book can be obtained by:

- downloading from the IET Digital Library at https://digital-library.theiet.org/content/books/ra/pbpo118e. Once on the page, click on the "Supplementary Material" tab to access the files.
- alternatively email books@theiet.org to be sent a copy of the files.

Chapter 2

Digital simulation

Andrea Benigni[1,2]

The main purpose of this chapter is to review some basic aspect of numerical integration of differential equations. Implicit and explicit as well as single and multistep algorithms have been considered. References [1–3] are used for this paragraph. References [1,2] are very well-known texts about numerical integration of differential equations and can be used for a deep understanding of the math behind what is presented in this chapter, Ref. [3] is a simulation textbook that—in the first chapter—offers a very well done and comprehensive review of integration methods applied to simulation of continuous system.

Let us start looking to what a differential equation is. An equation that involves one or more derivatives of the unknown function is called an ordinary differential equation, abbreviated as ODE. The order of the equation is determined by the order of the highest derivative, e.g. if the first derivative is the only derivative, then the equation is called a first-order ODE.

The solution of ODEs is classified in two groups of problems: the *initial-value problems* (IVP) and *boundary-value problems* (BVP). All the conditions of an IVP are specified at the initial point. On the other hand, the problem becomes a BVP if the conditions are needed for both initial and final points. The solution of ODEs in time domain is typically an IVP since all conditions are specified for the initial time.

The modeling of many engineering problems gives rise to a system of ODEs; as a consequence, the solution of ODEs and especially of the IVP has great importance in many practical applications. Unfortunately, at the same time, the number of cases in which the exact solution for an ODE can be found analytically is really limited. In this context, stable and accurate numerical methods that can approximate the solution of any generic ODEs system are extremely critical.

Let us start defining a state-space model as in (2.1) with initial conditions $x(t_0) = x_0$.

$$\dot{x}(t) = f(x(t), u(t), t) \tag{2.1}$$

where $x(t)$ is the state variables vector. $u(t)$ is the input vector.

[1]Institute of Energy and Climate Research: Energy Systems Engineering (IEK-10), Juelich Research Center, Germany
[2]Department of Mechanical Engineering, RWTH Aachen University, Germany

Let be $x_i(t)$ a state trajectory of the system in (2.1). If f_i and all its derivatives are continuous function of time, $x_i(t)$ can be approximated, at any desired precision, by the Taylor expansion around any point of the trajectory.

The value of x_i can so be calculated at any point in time using (2.2):

$$x_i(T + \Delta t) = x_i(T) + f_i(T)\Delta t + \frac{df_i(T)}{dt}\frac{\Delta t^2}{2!} + \frac{d^2 f_i(T)}{dt^2}\frac{\Delta t^3}{3!} + \ldots \qquad (2.2)$$

It is easy to imagine that given an initial condition and imposed $T = t_0$, the value of x_i can be computed for any value of Δt. Nevertheless, in practical simulation, equation (2.2) is hardly directly computed due to the complexity and the cost of analytically calculate the derivatives of x_i. At the same time, (2.2) has a strong theoretical value and integration methods can be classified on the base of how many terms of (2.2) they approximate.

Numerical methods can be, first of all, classified in explicit and implicit methods. Explicit methods directly calculate the state of a system at the later time $T + \Delta t$ from the state of the system at time T as in (2.3):

$$x(T + \Delta t) = f(x(T), u(T), \Delta t) \qquad (2.3)$$

In implicit methods, the calculation of the state of a system at the later time $T + \Delta t$ depends from the state of the system at the same time (2.4):

$$x(T + \Delta t) = f(x(T), x(T + \Delta t), u(T + \Delta t), \Delta t) \qquad (2.4)$$

The main drawback of implicit methods is the additional computational cost that derives from the required simultaneous solution of all the equations of the considered system; this may require at least one matrix inversion per simulation step (assuming a generic non-linear system). At the same time, implicit methods typically have a very good stability property and have been widely used for the simulation of electronics circuit and in general for stiff problems. Explicit methods vice versa have, typically, a smaller computational cost but, at the same time, they require—usually—a smaller time step to be stable and accurate.

Another important classification for integration methods has to be done between single- and multi-step methods. Single-step methods are based on the previous solution of the state and, depending on the specific method used, they may require multiple derivative evaluations. Multistep methods, with the purpose of limiting the number of derivatives evaluations, use multiple previous steps. Single-step methods are more diffused as a consequence of their simplicity of implementation, nevertheless, multistep methods have to be in general considered as a valid alternative due to the significant computational cost reduction they may offer.

In the rest of this chapter, an overview of various integration algorithms is presented. Implicit and explicit as well as single- and multi-step algorithms have been considered.

Before proceeding, let us make a generic consideration on the choice of the integration step size. The integration step size has a major impact on simulation accuracy and stability and, as consequence, it has to be defined small enough to capture all the relevant dynamics but long enough to limit the calculation time impact on the process itself. This consideration has to be kept in mind when comparing integration methods.

2.1   Euler forward method

The Euler methods are simple methods of solving first-order ODE, particularly suitable for quick programming because of their great simplicity. Let us consider the system in (2.1) and let us assume that t_c is the current time and $x(t_c)$ is the value of current state, our goal is to calculate the value of the future state $x(t_c + \Delta t)$. For a small enough Δt, the derivatives of $x(t)$ at t_c can be approximated by the forward difference as in (2.5):

$$\dot{x}(t_c) \approx \frac{x(t_c + \Delta t) - x(t_c)}{\Delta t} \tag{2.5}$$

This implies:

$$x(t_c + \Delta t) \approx x(t_c) + \Delta t\, \dot{x}(t_c) \tag{2.6}$$

By ignoring the approximation, we get:

$$x(t_c + \Delta t) = x(t_c) + \Delta t f(t_c, x(t_c), u(t)) \tag{2.7}$$

Looking at (2.7), it is clear that the obtained method is explicit, and, comparing (2.7) with (2.2), it can be easily seen that this method is equivalent to the Taylor expansion truncated after the second term. This method is usually referred as the Euler forward method.

Let us now consider the linear-time invariant differential equation in (2.8):

$$\dot{x} = \lambda x \tag{2.8}$$

Applying the Euler forward integration method, we obtain:

$$x(t_c + \Delta t) = (1 + \Delta t\, \lambda) x(t_c) \tag{2.9}$$

As a consequence, the integration is stable only if:

$$|1 + \Delta t \lambda| \leq 1 \tag{2.10}$$

From (2.10), the stability domain of the Euler forward method can be drawn in the complex plane $\Delta t \lambda$ (Figure 2.1).

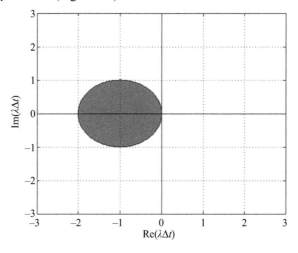

Figure 2.1   Stability domain of the Euler forward integration method

2.2   Backward Euler method

Let us now still consider the system in (2.1) but let us approximate the derivatives of $x(t)$ at t_c by the backward difference as in (2.11):

$$\dot{x}(t_c + \Delta t) \approx \frac{x(t_c + \Delta t) - x(t_c)}{\Delta t} \tag{2.11}$$

Rearranging the terms, we get:

$$x(t_c + \Delta t) \approx x(t_c) + \Delta t \dot{x}(t_c + \Delta t) \tag{2.12}$$

By ignoring the approximation:

$$x(t_c + \Delta t) = x(t_c) + \Delta t f(t_c + \Delta t, x(t_c + \Delta t), u(t + \Delta t)) \tag{2.13}$$

The integration method obtained in (2.13) is known as the backward Euler method and, as it can be seen, is an implicit method.

Applying Euler backward integration method to the equation in (2.8), (2.14) is obtained

$$x(t_c + \Delta t) = \left(\frac{1}{1 - \Delta t \, \lambda} \right) x(t_c) \tag{2.14}$$

As a consequence, the integration using the Euler backward method will be stable only if:

$$\left| \frac{1}{1 - \Delta t \, \lambda} \right| \leq 1 \tag{2.15}$$

From (2.15), the stability domain of the Euler backward method can be drawn in the complex plane $\Delta t \, \lambda$ (Figure 2.2).

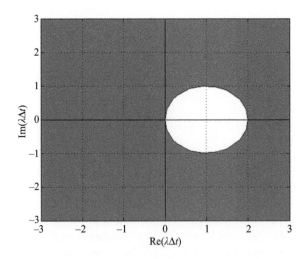

Figure 2.2   Stability domain of the Euler backward integration method

Table 2.1   Example 1.1 parameters

λ_1	-0.1
λ_2	-1
λ_3	-2
λ_4	-3

Example 1.1 In this example, we consider the equation of (2.8) and we show how the Euler forward and Euler backward integration methods behave for various values of λ. Let us assume that we fix $\Delta t = 1$ and we consider the four values of Table 2.1 for λ. The results of the integration performed using the Euler forward method are reported as in Figure 2.3 for each of the four considered values of λ, results are compared with the related analytical solution. The results of the integration performed using the Euler backward method are reported as in Figure 2.4.

Figure 2.3 clearly shows how changing the dynamic behavior of the system, making it "faster," reduces the accuracy of the Euler forward method even making the simulation unstable. In the case of Figure 2.3(a), where the time step is one order of magnitude smaller that the system dynamic, the simulation is stable and quite accurate; in Figure 2.3(b), the Euler forward method shows still a stable behavior, but the accuracy get strongly compromised suggesting that either the time step or the integration method should be reconsidered. It is worth to remember that how accurate an integration process has to be depends on the specific application. In general, an accuracy like the one of Figure 2.3(b) may be sufficient for a specific application but, at the same time, an accuracy like the one in Figure 2.3(a) may be completely unacceptable for another application.

It is interesting to notice how in Figure 2.3(c), the integration presents undamped oscillations, looking at the value λ we can see that we are on the stability border where $|1 + \Delta t \, \lambda| = 1$ (marginally stable). In Figure 2.3(d), we can be seeing how, being outside the stability region of the Euler forward method, the simulation gets unstable and diverge.

Let us now look at the results obtained using the Euler backward method (Figure 2.4). The first significant observation is that independently of the magnitude of λ, the integration is always stable. Indeed, an important property of the Euler backward method is the large stability region, as shown in Figure 2.2. At the same time, the large stability domain of the Euler backward method has also an inconvenience that does not make of the Euler backward always a suitable solution: since the Euler backward method can make the integration of an unstable system (by nature) stable, this may lead to serious simulation artifacts that may bring to wrong critical decision. This problem has to be taken in serious consideration when dealing with the system that may present unstable behavior.

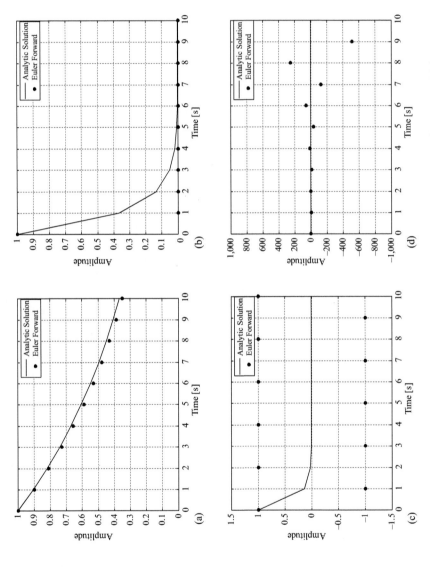

Figure 2.3 Integration of (2.8) with the Euler forward method ($\Delta t = 1$). (a) $\lambda_1 = -0.1$, (b) $\lambda_1 = -1$, (c) $\lambda_1 = -2$, and (d) $\lambda_1 = -3$.

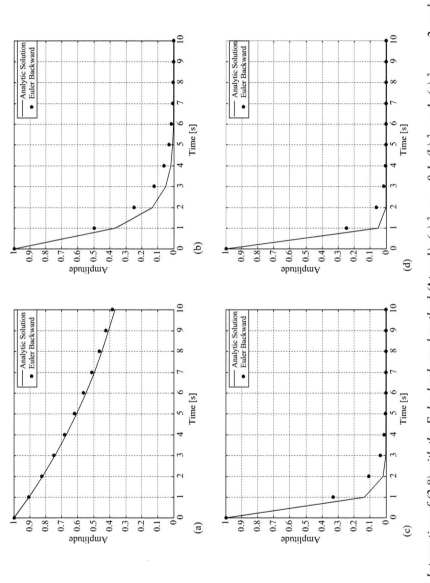

Figure 2.4 Integration of (2.8) with the Euler backward method ($\Delta t = 1$), (a) $\lambda_1 = -0.1$, (b) $\lambda_1 = 1$, (c) $\lambda_1 = -2$, and (d) $\lambda_1 = -3$.

2.3   Trapezoidal rule method

Let us now consider a different integration method largely known with the name of trapezoidal rule. In this case, the calculation of the area between the curves we want to integrate and the $y = 0$ axis is linearly approximated.

Equation (2.16) presents the integration scheme for the trapezoidal rule.

$$x(t_c + \Delta t) = x(t_c) + \frac{\Delta t}{2} (f(t_c + \Delta t, x(t_c + \Delta t), u(t + \Delta t)) + f(t_c, x(t_c), u(t)))$$

(2.16)

The trapezoidal rule integration method is a second-order single-step method, it is a very well-known implicit method that presents two extremely interesting properties that made its fortune. The trapezoidal rule is stable, compared with Euler backward method; trapezoidal rule does not present the inconvenient of making unstable system in nature stable in simulation like Euler backward method: its stability domain is limited to the left half of the complex plane $\Delta t \, \lambda$. It can be proven that trapezoidal rule approximates the Taylor expansion up the second order.

Applying the trapezoidal rule to the equation in (2.8), (2.17) is obtained

$$x(t_c + \Delta t) = x(t_c) + \frac{\Delta t}{2} \lambda x(t_c) + \frac{\Delta t}{2} \lambda x(t_c + \Delta t)$$

(2.17)

Regrouping the terms, we get

$$x(t_c + \Delta t) = \frac{1 + \lambda \frac{\Delta t}{2}}{1 - \lambda \frac{\Delta t}{2}} x(t_c)$$

(2.18)

The trapezoid rule method is stable if

$$\left| \frac{1 + \lambda \frac{\Delta t}{2}}{1 - \lambda \frac{\Delta t}{2}} \right| \leq 1$$

(2.19)

As anticipated before, the stability domain of trapezoidal rule includes all and only the left half plane of $\Delta t \, \lambda$.

To calculate the integral more accurately, the interval Δt can be divided into n segments each of size h. The trapezoidal rule is then applied to each segment. The sum of the results obtained for each segment is the approximate value of the integral. So, the integral can be broken into h integrals:

$$\int_{t_c}^{t_c + \Delta t} \dot{x} dx = \int_{t_c}^{t_c + h} \dot{x} dx + \int_{t_c + h}^{t_c + 2h} \dot{x} dx + \ldots + \int_{t_c + (n-2)h}^{t_c + (n-1)h} \dot{x} dx + \int_{t_c + (n-1)h}^{t_c + \Delta t} \dot{x} dx$$

(2.20)

This approach takes the name of multiple segment trapezoidal rule:

$$x(t_c + \Delta t) = x(t_c) + \frac{\Delta t}{2n} \left[f(t_c, x(t_c), u(t)) \right.$$

$$+ 2 \left\{ \sum_{i=1}^{n-1} f\left(t_c + ih, x(t_c + ih), u(t + ih) \right) \right\}$$

$$\left. + f(t_c + \Delta t, x(t_c + \Delta t), u(t + \Delta t)) \right] \qquad (2.21)$$

2.4   Predictor and corrector method

The predictor and corrector integration method is based on two evaluations of the derivative. The predictor, using an explicit method, calculates a rough estimation of the value for the next step of the function to be integrated. The corrector, using an implicit method, calculates the final value for the next step of the function to be integrated on the base of the value computed by the predictor. It is important to clarify immediately that such a predictor and corrector scheme has to be classified as an explicit method.

A very simple predictor and corrector can be obtained using the Euler forward method for the predictor and the Euler backward method for the corrector. Applying such a technique to the system in (2.1), we get:

$$x^P(t_c + \Delta t) = x(t_c) + \Delta t f(t_c, x(t_c), u(t)) \qquad (2.22)$$

x^P is the predicted value for x obtained using the Euler forward method. We can evaluate the derivative using $x^P(t_c + \Delta t)$ and use it in the corrector step based on the Euler backward method:

$$x(t_c + \Delta t) = x(t_c) + \Delta t f\left(t_c + \Delta t, x^P(t_c + \Delta t), u(t + \Delta t) \right) \qquad (2.23)$$

It is easy to imagine that the value just calculated with (2.23) could be reused to perform another correction step and that this process can be infinitely repeated. This process usually known as the Newton iteration is typically used for the non-linear system for which the direct use of an implicit method would not be convenient. The process of correction is iterated until two consecutive approximations of $x(t_c + \Delta t)$ differ less than a prescribed tolerance.

While the predictor and corrector scheme presented can significantly increase the accuracy of the integration, if compared to the simple explicit method used for the prediction, unfortunately the stability region is even reduced by this method.

2.5   Runge–Kutta methods

An important family of explicit method is the one of the Runge–Kutta methods. The Euler forward method is often inefficient because of the small step size required to obtain a specified accuracy, furthermore, round-off error accumulation, when many steps are used, can make the numerical results unusable. For these reasons, there always been interest in developing integration algorithms that still presenting an explicit behavior offer better accuracy than Euler forward method. In this chapter, it is not reported how the different Runge–Kutta methods are derived, the integration schemes of the considered orders are simply listed.

The first-order Runge–Kutta method, often abbreviated in RK1, is equivalent to Euler forward (2.13). The second-order Runge–Kutta method, RK2, also known as explicit midpoint rule, can be interpreted as a two-stage predictor and corrector. The integration scheme for RK2 is here reported in (2.24)–(2.27)

$$k_1 = f(t_c, x(t_c), u(t)) \tag{2.24}$$

$$x\left(t_c + \frac{\Delta t}{2}\right) = x(t_c) + \frac{\Delta t}{2} k_1 \tag{2.25}$$

$$k_2 = f\left(t_c + \frac{\Delta t}{2}, x\left(t_c + \frac{\Delta t}{2}\right), u\left(t_c + \frac{\Delta t}{2}\right)\right) \tag{2.26}$$

$$x(t_c + \Delta t) = x(t_c) + \Delta t k_2 \tag{2.27}$$

This symmetrization cancels out the first-order error of RK1, creating so a second-order method. By using two intermediate steps per interval, it is possible to cancel out both the first and second-order error terms, and, so obtain a third-order Runge–Kutta method, RK3. In the same way, if three intermediate steps are considered, the most often used fourth-order Runge–Kutta method (RK4) can be derived. The integration scheme for RK4 is here reported in (2.28)–(2.32):

$$k_1 = f(t_c, x(t_c), u(t)) \tag{2.28}$$

$$k_2 = f\left(t_c + \frac{\Delta t}{2}, x(t_c) + \frac{\Delta t}{2} k_1, u\left(t_c + \frac{\Delta t}{2}\right)\right) \tag{2.29}$$

$$k_3 = f\left(t_c + \frac{\Delta t}{2}, x(t_c) + \frac{\Delta t}{2} k_2, u\left(t_c + \frac{\Delta t}{2}\right)\right) \tag{2.30}$$

$$k_4 = f(t_c + \Delta t, x(t_c) + \Delta t k_3, u(t_c + \Delta t)) \tag{2.31}$$

$$x(t_c + \Delta t) = x(t_c) + \frac{\Delta t k_1}{6} + \frac{\Delta t k_2}{3} + \frac{\Delta t k_3}{3} + \frac{\Delta t k_4}{6} \tag{2.32}$$

Applying the different Runge–Kutta methods to the problem of (2.8), (2.33)–(2.36) are so obtained:

$$\text{RK1}: x(t_c + \Delta t) = (1 + \Delta t \lambda) x(t_c) \tag{2.33}$$

$$\text{RK2} : x(t_c + \Delta t) = \left(1 + \Delta t \lambda + \frac{(\Delta t \lambda)^2}{2}\right) x(t_c) \tag{2.34}$$

$$\text{RK3} : x(t_c + \Delta t) = \left(1 + \Delta t \lambda + \frac{(\Delta t \lambda)^2}{2} + \frac{(\Delta t \lambda)^3}{6}\right) x(t_c) \tag{2.35}$$

$$\text{RK4} : x(t_c + \Delta t) = \left(1 + \Delta t \lambda + \frac{(\Delta t \lambda)^2}{2} + \frac{(\Delta t \lambda)^3}{6} + \frac{(\Delta t \lambda)^4}{24}\right) x(t_c) \tag{2.36}$$

Equations from (2.33) to (2.36) can so be used to define the regions of absolute stability for each Runge–Kutta method (2.37)–(2.40):

$$\text{RK1} : |1 + \Delta t \lambda| < 1 \tag{2.37}$$

$$\text{RK2} : \left|1 + \Delta t \lambda + \frac{\Delta t \lambda^2}{2}\right| < 1 \tag{2.38}$$

$$\text{RK3} : \left|1 + \Delta t \lambda + \frac{\Delta t \lambda^2}{2} + \frac{\Delta t \lambda^3}{6}\right| < 1 \tag{2.39}$$

$$\text{RK4} : \left|1 + \Delta t \lambda + \frac{\Delta t \lambda^2}{2} + \frac{\Delta t \lambda^3}{6} + \frac{\Delta t \lambda^4}{24}\right| < 1 \tag{2.40}$$

In Figure 2.5, the stability domain for the four considered Runge–Kutta algorithms is reported. Two important considerations should be done. High-order Runge–Kutta algorithms present a significantly bigger region of stability if compared with the Euler forward method. One important property of the fourth-order

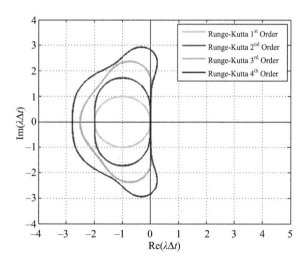

Figure 2.5   Stability domain of different order Runge–Kutta integration method

Runge–Kutta algorithm, as can be observed in Figure 2.5, is that allows stable simulation of system with purely imaginary eigenvalues.

2.6   Adams–Bashforth and Adams–Moulton methods

To complete this overview of integration methods, an explicit and an implicit multistep integration method are presented. As in the case of Runge–Kutta methods, only the integration schemes are presented.

In (2.41)–(2.43), the second-, third-, and fourth-order explicit Adams–Bashforth integration methods are reported. The first-order Adams–Bashforth method is not reported here since it is equivalent to the Euler forward method:

$AB2 : x(t_c + \Delta t)$

$$= x(t_c) + \Delta t \left[\frac{3}{2} f(t_c, x(t_c), u(t)) - \frac{1}{2} f(t_c - \Delta t, x(t_c - \Delta t), u(t - \Delta t)) \right]$$

$$(2.41)$$

$AB3 : x(t_c + \Delta t) = x(t_c)$

$$+ \Delta t \left[\frac{23}{12} f(t_c, x(t_c), u(t)) - \frac{16}{12} f(t_c - \Delta t, x(t_c - \Delta t), u(t - \Delta t)) \right.$$

$$\left. + \frac{5}{12} f(t_c - 2\Delta t, x(t_c - 2\Delta t), u(t - 2\Delta t)) \right]$$

$$(2.42)$$

$AB4 : x(t_c + \Delta t) = x(t_c)$

$$+ \Delta t \left[\frac{55}{24} f(t_c, x(t_c), u(t)) - \frac{59}{24} f(t_c - \Delta t, x(t_c - \Delta t), u(t - \Delta t)) \right.$$

$$+ \frac{37}{24} f(t_c - 2\Delta t, x(t_c - 2\Delta t), u(t - 2\Delta t))$$

$$\left. - \frac{9}{24} f(t_c - 3\Delta t, x(t_c - 3\Delta t), u(t - 3\Delta t)) \right]$$

$$(2.43)$$

In Figure 2.6, the stability domain for Adams–Bashforth methods from first to fourth order is reported. As shown in Figure 2.6, the pretty disappointing result is that increasing the order of integration, we reduce the size of the stability region.

We can conclude that if on the one side Adams–Bashforth method are less computationally costly, since only one function evaluation is required, probably a much smaller time step will be required. As a consequence, the selection of a Adams–Bashforth integration method instead of a Runge–Kutta one should be done after accurate evaluation of the system to be integrated.

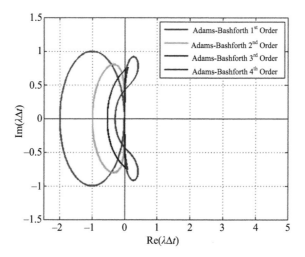

Figure 2.6   Stability domain of different order Adams–Bashforth integration method

A simple example of the implicit multi-step method is represented by the Adams–Moulton methods. First- and second-order methods are not reported here since they coincide with the Euler backward method and the trapezoidal rule, respectively:

$$\text{AM3}: x(t_c + \Delta t) = x(t_c) + \Delta t \left[\frac{5}{12} f(t_c + \Delta t, x(t_c + \Delta t), u(t + \Delta t)) \right.$$

$$\left. + \frac{86}{12} f(t_c, x(t_c), u(t)) - \frac{1}{12} f(t_c - \Delta t, x(t_c - \Delta t), u(t - \Delta t)) \right] \qquad (2.44)$$

$$\text{AM4}: x(t_c + \Delta t) = x(t_c) + \Delta t \left[\frac{9}{24} f(t_c + \Delta t, x(t_c + \Delta t), u(t + \Delta t)) \right.$$

$$+ \frac{19}{24} f(t_c, x(t_c), u(t)) - \frac{5}{24} f(t_c - \Delta t, x(t_c - \Delta t), u(t - \Delta t))$$

$$\left. + \frac{1}{24} f(t_c - 2\Delta t, x(t_c - 2\Delta t), u(t - 2\Delta t)) \right] \qquad (2.45)$$

In Figure 2.7, the stability domain for the considered Adams–Moulton algorithms is reported. Even if the stability regions presented in Figures 2.6 and 2.7 do not seem to encourage the use of these methods, Adams–Bashforth and Adams–Moulton methods can be combined together to define very stable predictor and corrector schemes.

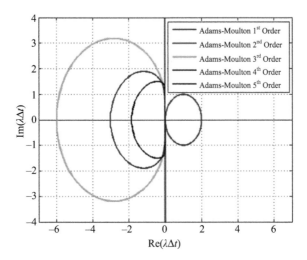

Figure 2.7   Stability domain of different order Adams–Moulton integration method

2.7   Accuracy comparison

In this last paragraph dedicated to integration methods, a simple comparison of the accuracy offered by different integration methods is presented:

The characteristic equation of an inductor is taken as the reference case. The Z-domain transfer function in the form of (2.45) has been created for each single method taken into consideration.

$$H(z) = \frac{I(z)}{V(z)} \tag{2.46}$$

Comparing the frequency response obtained with the discrete time transfer function with the one obtained by the continuous time transfer function, results in Figure 2.8(a) and (b) are obtained. It is worth to underline that the frequency axis is labeled in p.u. of $1/dt$ up to 0.5 that corresponds to the Nyquist frequency. An interesting aspect to notice is that trapezoidal rule not only has a really convenient region of stability, as it is already been shown, but it is also extremely accurate if compared to other integration methods of similar order.

Exercises

Exercise 1 Consider the following initial value problem:

$$\dot{x}(t) = -5x(t) + 10$$
$$x(0) = 4$$

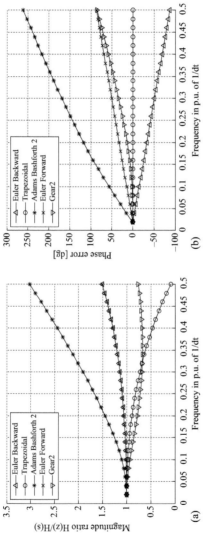

Figure 2.8 Accuracy comparison: (a) magnitude and (b) phase

Assuming initial time $t_{init} = 0\,\text{s}$, and time step $\Delta t = 0.1\,\text{s}$

1. Is the system described by the differential equation above stable? What if $x(0) = -8$
2. Compute two simulation steps (calculate $x(0.1\,\text{s})$ and $x(0.2\,\text{s})$) using the Euler forward method, the Euler backward method, and the trapezoidal rule.
3. Assuming $\Delta t = 1\,\text{s}$, is the numerical integration stable? Compute two simulation steps (calculate $x(1\,\text{s})$ and $x(2\,\text{s})$) using the Euler forward method, the Euler backward method, and the trapezoidal rule.

Exercise 2

Consider the following initial value problem:

$$\dot{x}(t) = -10x(t) + 11$$
$$x(0) = 3$$

1. Is the system described by the differential equation above stable?
2. Assume to use the Euler forward as integration method. Calculate the time step size that will results in a marginally stable solution.

Exercise 3

Consider the following initial value problem:

$$\dot{x}(t) = 10x(t) + 11$$
$$x(0) = 3$$

1. Is the system described by the differential equation above stable?
2. Assume to use the Euler backward as an integration method and $\Delta t = 1\,\text{s}$, is the numerical integration stable?

References

[1] J.D. Lambert, *"Numerical Methods for Ordinary Differential Systems, The Initial Value Problem"*, John Wiley & Sons Ltd., New York, NY, 1991.
[2] C.W. Gear, *"Numerical Initial Value Problems in Ordinary Differential Equations"*, Prentice-Hall Inc., Englewood Cliffs, NJ, 1971.
[3] F.E. Cellier and E. Kofman, *"Continuous System Simulation"*, Springer, New York, NY, 2006, ISBN 0-387-26102-8.

Chapter 3

Nodal methods

Antonello Monti[1,2] and Andrea Benigni[3,4]

A method that historically played a leading role for the modeling of electric circuit for transient simulation is the resistive companion [1] also often known as Dommel algorithm [2].

We now proceed presenting the principles of nodal analysis [3] and modified nodal analysis [4] that are on the basis of the resistive companion algorithm. Nodal analysis and modified nodal analysis are first presented with reference to DC circuit, immediately after is shown how, with resistive companion, nodal analysis and modified nodal analysis are used to solve circuit containing dynamic linear elements. To conclude how resistive companion can be extended to solve the more general family of non-linear dynamic network is illustrated.

Let us suppose we have a circuit composed by b branches. The solution of this circuit is fully accomplished if we determine the voltage and current in each branch. As a result, the problem concerns with the calculation of $2b$ unknowns. Linear algebra tells us that in order to properly solve this problem we need $2b$ independent equations.

Let us focus on a simple linear system. The equations we have available are:

- Kirchhoff current laws (KCL)
- Kirchhoff voltage laws (KCV)
- Characteristic equations (Ohm's Law in a broad sense).

It is easy to see that we can write one characteristic equation for each branch of the network, for a total number of b equations. Assuming then that the topology of the network is characterized by n nodes, it is to prove that we can write $n-1$ independent KCL. The nth equation can be easily shown being a linear combination of the remaining $n-1$.

At the same time, it can be demonstrated that if we can imagine a very high number of closed path in the network only $b-n+1$ are able to provide independent KVL.

To prove this statement, let us consider a generic electrical circuit, like the one of Figure 3.1, and let us now substitute the circuit with on oriented graph i.e. with a

[1]Institute for Automation of Complex Power Systems, RWTH Aachen University, Germany
[2]Fraunhofer FIT Center for Digital Energy, Germany
[3]Institute of Energy and Climate Research: Energy Systems Engineering (IEK-10), Juelich Research Center, Germany
[4]Department of Mechanical Engineering, RWTH Aachen University, Germany

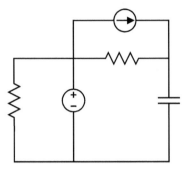

Figure 3.1   A simple electric circuit

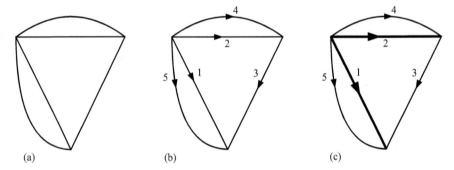

*Figure 3.2   (a) The graph representing the topology of the electrical circuit of
previous figure, (b) the oriented version of the graph, and (c) the
oriented graph with selected a possible tree*

diagram composed of oriented arcs where each arc represents one of the branches
of the original circuit. With this substitution, we can focus simply on the topology
disregarding the characteristics of the specific branch. In effect, our conclusion
regarding the KVL must be valid for every network no matter the branches
involved and then it makes sense to focus only on the connections.

Let us suppose for example that the diagram looks like in Figure 3.2(a). Let us
now orient the arc, exactly as we assume a direction for the measurement of the
current and let us also number the arcs, Figure 3.2(b). We can now define a tree for
the oriented graph. A tree is given by a subset of arcs that connects all the nodes but
do not create any closed path.

To connect n nodes without creating a closed path we are going to need $n-1$
arcs. As a result, $b-n+1$ arcs are not included in the tree (this is actually called co-
tree). For this network for example we have $b = 5$ and $n = 3$ so the tree includes
two branches and the co-tree includes three branches.

It is pretty easy to see that we can write one independent KVL for each branch
in the co-tree. These are linearly independent because each of them contains at least
one branch that does not appear in any of the other equations. In this case we have
$b-n+1$ independent KVL as in (3.1), (3.2), and (3.3).

$$v_1 + v_1 - v_2 = 0 \tag{3.1}$$

$$v_2 - v_4 = 0 \tag{3.2}$$

$$v_5 - v_1 = 0 \tag{3.3}$$

It is not possible to write any new independent equation, as any other equation we will write can be obtained as linear combination of the one already written. The linear combination can be determined looking at which branches of the co-tree are included.

If for example we consider (3.4), this is not independent. It involves the branches of the co-tree 3 and 4 and then it can be obtained combining the equations of the original set that contain these two branches:

$$v_1 + v_3 - v_4 = 0 \tag{3.4}$$

In our case, we can easily see in effect that (3.4) is the sum of (3.1) and (3.2). The same can be shown for any other closed path.

If we put everything together, we have $n-1$ KCL, $b-n+1$ KVL and b characteristic equations. In total this means we have 2b equations. As a consequence the solution of a linear network is a well posed problem with a unique solution. Notice that the linearity only affects the characteristic equations, while for any network (linear/non-linear, time variant or time invariant) the b equations defined by the Kirchhoff equations always hold.

3.1 Nodal analysis

Now that we have demonstrated that the problem can be solved, we want to determine a smart way to approach it. The idea is to find an easy approach that will allow a reduction of the number of equations we need to solve simultaneously. Traditionally two methods have been proposed in literature:

- Mesh current analysis
- Nodal analysis

In the case of the mesh current analysis, we identify as unknowns the currents that we assume are flowing through a set of independent meshes. Thanks to the KCL, we can relate these currents to the branch currents and thanks to Ohm's Law we can remove the voltage from all the KVL and express everything as function of the currents. As result of that we obtain a set of $b-n+1$ equation that we have to solve simultaneously.

Nodal analysis [3] is the dual approach and it is based on the selection of $n-1$ independent voltages. An example of such set is given by the voltage (potential) of each node with respect to a node selected as reference. Exactly as in the case of the mesh analysis, the KVL will allow us to relate the branch voltages to the node potential and Ohm's Law will allow us to remove the currents from every KCL that we write. In this case, we have to solve $n-1$ independent equations.

Historically the second option has been considered the best option to adopt when designing a circuit simulator. Some reasons can be found to justify this choice:

- Most generators behave as voltage sources and then the determination of the voltage profile is a point of major interest.

- Nodal analysis allows for an easy definition of an open circuit while cannot describe a perfect short circuit. The opposite is true for mesh analysis. In reality, short circuits never convey zero resistance and then the limitation of nodal analysis is meaningless.
- In electronic circuit, the information is fully contained in the voltage and then this is the main quantity we want to solve for.

In any case, the application of one of the two methods brings a significant benefit to the circuit analysis. It should be reminded that the solution of k simultaneous equation can be formulated in matrix form as the solution of the linear problem (3.5)

$$Gx = b \qquad (3.5)$$

where A is the k size square matrix of known coefficients, x is a vector of dimension k of unknowns and b is a vector of dimension k of known coefficients. The solution of such problem is obtained inverting the matrix A:

$$x = G^{-1}b \qquad (3.6)$$

In reality, the matrix G is never fully inverted, but numerical methods that directly allow the calculation of vector x are used; one of the most common is the method which we will refer to for the rest of this dissertation as the LU factorization. Computationally speaking, the solution of the linear problem of (3.5) is a very intense problem with a complexity that is approximately proportional to the third power of the size of the system. As results of this, the reduction of the number of unknowns has a very significant impact on the complexity of the problem.

Example 2.1 Let us solve this circuit of Figure 3.3 by using nodal analysis. One node is defined as reference and then if we consider as possible branches resistors and voltage sources with resistance in series we have to determine the potential for two independent nodes. We can then write two KCL: (3.7) and (3.8):

$$\frac{e_1 - V_1}{R_1} + \frac{e_1}{R_2} + \frac{e_1 - e_2}{R_3} = 0 \qquad (3.7)$$

$$\frac{e_2 - V_5}{R_5} + \frac{e_2}{R_4} + \frac{e_2 - e_1}{R_3} = 0 \qquad (3.8)$$

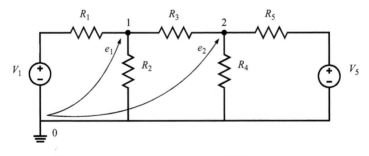

Figure 3.3   *A simple example of DC linear circuit*

Rearranging the terms and putting the solution in the matrix form, we obtain:

$$\begin{bmatrix} \dfrac{1}{R_1} + \dfrac{1}{R_2} + \dfrac{1}{R_3} & -\dfrac{1}{R_3} \\ -\dfrac{1}{R_3} & \dfrac{1}{R_3} + \dfrac{1}{R_4} + \dfrac{1}{R_5} \end{bmatrix} \begin{bmatrix} e_1 \\ e_2 \end{bmatrix} = \begin{bmatrix} \dfrac{V_1}{R_1} \\ \dfrac{V_5}{R_5} \end{bmatrix} \tag{3.9}$$

Solving the system of (3.9), we can compute the voltage potentials. Since no time variant elements are considered the solution is constant over time as in Figure 3.4.

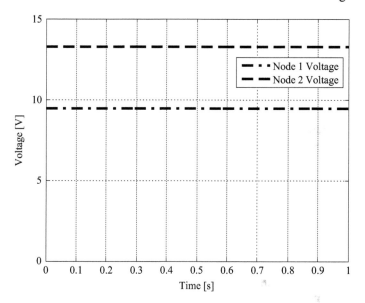

Figure 3.4 Example 1.1 solution

3.2 Matrix stamp

Observing carefully the structure of the matrices obtained for the solution of the previous example we can identify a general approach to the construction of the matrix A and the vector b. This concept is called matrix stamp because the basic idea is that each component stamps with its presence certain positions in the matrices structure. Matrices are initialized with all zeros on the base of the number of node in the circuit. We will now proceed looking at some of the most common components that can be treated with nodal analysis. From now on, we will refer to G as conductance matrix and to b as source vector.

3.2.1 Resistor

Let us suppose we have a resistor R_k between the nodes i and j (see Figure 3.5).

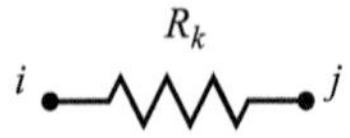

$$R_k$$

$$i \ \text{—}\bigwedge\bigwedge\text{—} \ j$$

Figure 3.5   Sample resistor between node i and j

The elements that will be affected are the elements of matrix A in the positions ii, ij, ji, jj. The stamp will be performed properly adding submatrix (3.10) to the conductance matrix, no values are added to the source vector:

$$
\begin{bmatrix}
\dfrac{1}{R_k} & -\dfrac{1}{R_k} \\[2mm]
-\dfrac{1}{R_k} & \dfrac{1}{R_k}
\end{bmatrix}
\begin{matrix} i \\[2mm] j \end{matrix}
\qquad (3.10)
$$

$$i \qquad j$$

3.2.2   Ideal current source

Let us suppose we have a current source I_k between the nodes i and j (see Figure 3.6).

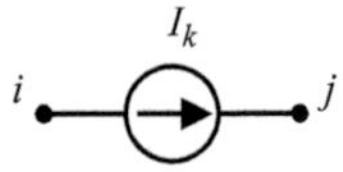

Figure 3.6   Sample ideal current source between node i and j

The elements that will be affected are the elements of the source vector in the positions i and j. The stamp will be performed properly adding vector (3.11) to the source vector; no elements are added to the conductance matrix:

$$
\begin{bmatrix} -I_k \\ I_k \end{bmatrix}
\begin{matrix} i \\ j \end{matrix}
\qquad (3.11)
$$

3.2.3   Real current source

Let us suppose we have a current source I_k between the nodes i and j with in parallel a resistance R_k (see Figure 3.7). The procedure is the composition of the matrix stamps of the resistor and of the ideal current source as reported above.

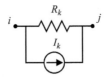

Figure 3.7   Sample real current source between node i and j

The elements that will be affected are the elements of the source vector in the positions i and j and the elements of the conductance matrix in the positions ii, jj, ij, ji. The stamp will be performed properly adding matrix (3.12) to the conductance matrix and vector (3.13) to the source vector:

$$
\begin{bmatrix}
\dfrac{1}{R_k} & -\dfrac{1}{R_k} \\[2mm]
-\dfrac{1}{R_k} & \dfrac{1}{R_k}
\end{bmatrix}
\begin{matrix} i \\[2mm] j \end{matrix}
\qquad (3.12)
$$

$$
\begin{bmatrix} -I_k \\ I_k \end{bmatrix}
\begin{matrix} i \\ j \end{matrix}
\qquad (3.13)
$$

3.2.4 Real voltage source

Let us suppose we have a voltage source V_k between the nodes i and j within series a resistance R_k (see Figure 3.8).

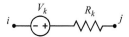

Figure 3.8 Sample real voltage source between node i and j

The way we perform matrix stamp can be understood assuming we perform first a transformation from Thevenin equivalent to Norton equivalent. The elements that will be affected are the elements of the source vector in the positions i and j and the elements of the conductance matrix in the positions ii, jj, ij, ji. The stamp will be performed properly adding matrix (3.14) to the conductance matrix and vector (3.15) to the source vector:

$$
\begin{bmatrix}
\dfrac{1}{R_k} & -\dfrac{1}{R_k} \\[2mm]
-\dfrac{1}{R_k} & \dfrac{1}{R_k}
\end{bmatrix}
\begin{matrix} i \\[2mm] j \end{matrix}
\qquad (3.14)
$$

$$
\begin{bmatrix} -\dfrac{V_k}{R_k} \\[3mm] \dfrac{V_k}{R_k} \end{bmatrix}
\begin{matrix} i \\[3mm] j \end{matrix}
\qquad (3.15)
$$

Example 2.2 Let us now solve the same circuit as in Figure 3.3, but using matrix stamp. The first step is the creation of the two matrices: the conductance matrix and the source vector. To properly initialize them we need to define the number of independent nodes. For this circuit, we have 2 independent nodes and so the conductance matrix is going to be a 2×2 matrix and the source vector is a 2×1.

$$\begin{bmatrix} 0 & 0 \\ 0 & 0 \end{bmatrix} \begin{bmatrix} e_1 \\ e_2 \end{bmatrix} = \begin{bmatrix} 0 \\ 0 \end{bmatrix} \qquad \begin{bmatrix} \frac{1}{R_1} & 0 \\ 0 & 0 \end{bmatrix} \begin{bmatrix} e_1 \\ e_2 \end{bmatrix} = \begin{bmatrix} \frac{V_1}{R_1} \\ 0 \end{bmatrix} \qquad \begin{bmatrix} \frac{1}{R_1} + \frac{1}{R_2} & 0 \\ 0 & 0 \end{bmatrix} \begin{bmatrix} e_1 \\ e_2 \end{bmatrix} = \begin{bmatrix} \frac{V_1}{R_1} \\ 0 \end{bmatrix}$$

(a)   (b)   (c)

$$\begin{bmatrix} \frac{1}{R_1} + \frac{1}{R_2} + \frac{1}{R_3} & -\frac{1}{R_3} \\ -\frac{1}{R_3} & \frac{1}{R_3} \end{bmatrix} \begin{bmatrix} e_1 \\ e_2 \end{bmatrix} = \begin{bmatrix} \frac{V_1}{R_1} \\ 0 \end{bmatrix} \qquad \begin{bmatrix} \frac{1}{R_1} + \frac{1}{R_2} + \frac{1}{R_3} & -\frac{1}{R_3} \\ -\frac{1}{R_3} & \frac{1}{R_3} + \frac{1}{R_4} \end{bmatrix} \begin{bmatrix} e_1 \\ e_2 \end{bmatrix} = \begin{bmatrix} \frac{V_1}{R_1} \\ 0 \end{bmatrix}$$

(d)   (e)

$$\begin{bmatrix} \frac{1}{R_1} + \frac{1}{R_2} + \frac{1}{R_3} & -\frac{1}{R_3} \\ -\frac{1}{R_3} & \frac{1}{R_3} + \frac{1}{R_4} + \frac{1}{R_5} \end{bmatrix} \begin{bmatrix} e_1 \\ e_2 \end{bmatrix} = \begin{bmatrix} \frac{V_1}{R_1} \\ \frac{V_5}{R_5} \end{bmatrix}$$

(f)

Figure 3.9   Example 1.1 solution using matrix stamp

At the first step, the matrices are initialized with zeros (Figure 3.9(a)). We can now proceed adding the real voltage source (V_1, R_1). This is between the node 0 and the node 1. We drop everything that regards the node 0 and then the matrix stamp will affect the element (1,1) of the conductance matrix and the element (1) of the source vector (Figure 3.9(b)). We move now to R_2. This is between node 1 and node 0. We drop everything that regards the node 0 and then the matrix stamp will affect the element (1,1) of the conductance matrix (Figure 3.9(c)). Let us proceed with R_3. This is between the node 1 and the node 2. The matrix stamp will affect the elements (1,1), (1,2), (2,1), and (2,2) of the conductance matrix (Figure 3.9(d)). R_4 is between the node 2 and the node 0. The matrix stamp will affect the element (2,2) of the conductance matrix (Figure 3.9(e)). As last, we look at the real voltage source (V_5, R_5). This is between the node 2 and the node 0. The matrix stamp will affect the element (2,2) of the conductance matrix and the element 2 of the source vector (Figure 3.9(f)).

3.3   Modified nodal analysis

In the previous paragraph, we saw how to perform matrix stamp for a family of components. Still the previous procedure cannot be extended to an ideal voltage source (see for example Figure 3.10). On the other hand, ideal voltage sources are important to represent real devices as controlled power supplies. The extension of nodal analysis to treat also ideal voltage sources take the name of modified nodal analysis [4]. This method consists in adding one equation and one unknown (the current through the source).

Figure 3.10 Sample ideal voltage source

Let us deduce how modified nodal analysis approach works by means of an example. In solving the circuit of Figure 3.11, we want to consider the voltage source V as an ideal voltage source. As result of that, the network has 3 nodes and then 2 independent nodes. The idea of the modified nodal analysis is to add one unknown which is the current through the source and one equation defined by the voltage across the source expressed as function of the node potentials.

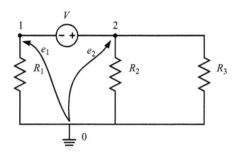

Figure 3.11 An example containing an ideal voltage source

Let us then write the equations manually first:

$$\frac{e_1}{R_1} - I = 0 \tag{3.16}$$

$$\frac{e_2}{R_2} + \frac{e_2}{R_3} + I = 0 \tag{3.17}$$

$$e_2 - e_1 = V \tag{3.18}$$

If we put it in matrix form, we have:

$$\begin{bmatrix} \dfrac{1}{R_1} & 0 & -1 \\ 0 & \dfrac{1}{R_2} + \dfrac{1}{R_3} & 1 \\ -1 & 1 & 0 \end{bmatrix} \begin{bmatrix} e_1 \\ e_2 \\ I \end{bmatrix} = \begin{bmatrix} 0 \\ 0 \\ V \end{bmatrix} \tag{3.19}$$

Analyzing matrix (3.19), we can deduce that the matrix stamp process can be modified as in the following:

1. Declare a conductance matrix of dimension $(n-1) \times (n-1)$, where n is the total number of nodes of the network and a source vector of dimension $(n-1) \times 1$.

2. Expand those matrices adding as many rows and columns to the conductance matrix as the number of ideal voltage sources; add as many rows as the number of the ideal voltage sources to the source vector.
3. For the kth ideal voltage source: stamp matrix (3.20) in the conductance matrix and vector (3.21) in the source vector:

$$
\begin{bmatrix}
 & & 1 \\
 & & -1 \\
1 & -1 &
\end{bmatrix}
\begin{matrix}
j \\ i \\ p
\end{matrix}
\qquad (3.20)
$$

$$
\begin{matrix} \\ \\ \end{matrix}
$$

$$
\begin{bmatrix}
\\
V_k
\end{bmatrix} p
\qquad (3.21)
$$

where $p=k+n-1$.

Example 2.3 In Figure 3.12, we have the same circuit of Figure 3.3 that now we will solve using modified nodal analysis. We now have 4 independent node and 2 ideal voltage source. The conductance matrix so have size (6,6) and the source vector (6,1). The system of equations in (3.22) is the nodal solution of the circuit in Figure 3.12 applying modified nodal analysis and matrix stamp:

$$
\begin{bmatrix}
\dfrac{1}{R_1} & -\dfrac{1}{R_1} & 0 & 0 & 1 & 0 \\[2mm]
-\dfrac{1}{R_1} & \dfrac{1}{R_1}+\dfrac{1}{R_2}+\dfrac{1}{R_3} & -\dfrac{1}{R_3} & 0 & 0 & 0 \\[2mm]
0 & -\dfrac{1}{R_3} & \dfrac{1}{R_3}+\dfrac{1}{R_4}+\dfrac{1}{R_5} & -\dfrac{1}{R_5} & 0 & 0 \\[2mm]
0 & 0 & -\dfrac{1}{R_5} & \dfrac{1}{R_5} & 0 & 1 \\[2mm]
1 & 0 & 0 & 0 & 0 & 0 \\[2mm]
0 & 0 & 0 & 1 & 0 & 0
\end{bmatrix}
\begin{bmatrix}
e_1 \\ e_2 \\ e_3 \\ e_4 \\ I_1 \\ I_5
\end{bmatrix}
=
\begin{bmatrix}
0 \\ 0 \\ 0 \\ 0 \\ V_1 \\ V_5
\end{bmatrix}
\qquad (3.22)
$$

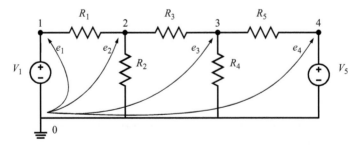

Figure 3.12   A simple example of DC linear circuit with ideal voltage source

Solving the system of (3.22), we can directly compute the four voltage potentials and the two currents in the ideal voltage sources. As for Example 2.1, since no time variant elements are considered, the solution is constant over time as in Figure 3.13. The same parameters used for Example 1.2 and reported in Table 3.2 have been used.

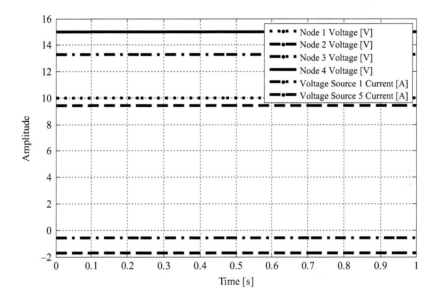

Figure 3.13   Example 1.3 solution

Table 3.1   Example 1.2 parameters

R_1,	R_5		1 Ω
R_2,	R_3,	R_4	10 Ω
V_1			10 V
V_5			15 V

Table 3.2   Example 2.5 parameters

R_1,	R_2	1 Ω
L		0.01 mH
C		0.01 mF
V		10 V

3.4   Resistive companion

The resistive companion method [1], also known as Dommel algorithm [2,5,6], is an approach to the solution of transient equation based on the transformation of every dynamic element in a correspondent DC equivalent circuit representing an iteration of an integration method.

Given a differential equation describing an electrical component, it has to be discretized so that can be expressed in the standard form of (3.23):

$$I(t + h) = g * V(t + h) - b(t) \tag{3.23}$$

where $I(t+h)$ is the current at the terminal of the dynamic component at next time step, $V(t+h)$ the voltage across the component at next time step, g a coefficient with conductance unit, and $b(t)$ an equivalent current representing the past history of the component. The relations between G, $b(t)$, the time step h, and the parameters of the device depend on the integration method selected for the process.

It is then clear that we can map the equation above in an equivalent circuit as in Figure 3.14.

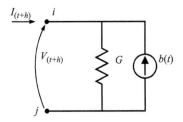

Figure 3.14   Standard components form for resistive companion

A generalized form of the conductance matrix and the source vector stamp when a dynamic component is represented using resistive companion is reported in (3.24) and (3.25):

$$\begin{bmatrix} g & -g \\ -g & g \end{bmatrix} \begin{matrix} i \\ j \end{matrix} \tag{3.24}$$
$$\begin{matrix} i & \quad j \end{matrix}$$

$$\begin{bmatrix} b(t) \\ -b(t) \end{bmatrix} \begin{matrix} i \\ j \end{matrix} \tag{3.25}$$

Example 2.4 Let us consider for example the case of a capacitor and discretize it using trapezoidal rule (see Figure 3.15).

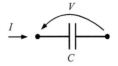

Figure 3.15   Capacitor

In (3.26), the voltage and the current relationship for the capacitor is reported. Rearranging (3.26) to isolate the derivative equation (3.27) is obtained:

$$i = C\frac{dv}{dt} \tag{3.26}$$

$$\frac{dv}{dt} = \frac{i}{C} \tag{3.27}$$

Applying trapezoidal rule to the characteristic equation of the capacitor, we obtain (3.28):

$$V(t+h) = V(t) + \frac{h}{2}\left(\frac{I(t+h)}{C} + \frac{I(t)}{C}\right) \tag{3.28}$$

In order to obtain the resistive companion form as in (3.23), (3.28) is rearranged so to isolate the current term for time t. Expression (3.29) is so obtained:

$$I(t+h) = \left(\frac{2C}{h}\right)V(t+h) - \left(\frac{2C}{h}\right)V(t) - I(t) \tag{3.29}$$

In (3.30) and (3.31), the conductance and state sources to be used in the matrix stamp process are reported:

$$g = \left(\frac{2C}{h}\right) \tag{3.30}$$

$$b(t) = \left(\frac{2C}{h}\right)V(t) + I(t) \tag{3.31}$$

3.4.1   Resistive companion solution flow

In Figure 3.16, resistive companion solution flow is depicted. The first step of the solution flow is the initialization. The initialization process depends on the specific algorithm adopted. Some algorithms may require for the initialization more information than what is typically provided as initial conditions.

In effect, initial conditions are usually provided in the form of voltage across the capacitors and current through the inductors. Some integration methods such as e.g. EF require also the evaluation of the other component variable: current through the capacitor and voltage across the inductors. This problem can be solved by performing the solution of the circuit at $t=0^+$ where the initial conditions are forced

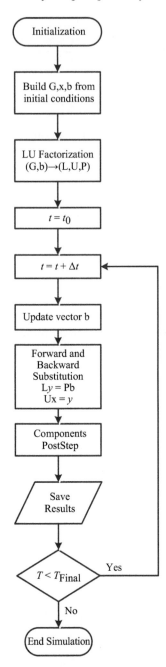

Figure 3.16   Resistive companion solution flow

and all the other variables are calculated. Notice that this solution is performed assuming that the time is frozen in a right neighborhood of $t=0$ forcing the initial conditions to be a continuous function.

Once the system is initialized, the conductance matrix G and the source vector b are created applying the matrix stamp approach. Since for now we are only considering linear systems, if we assume a constant time step for the whole simulation time, matrix G will never change; it so convenient to perform the LU factorization before entering the simulation loop so that it is not repeated at each iteration. We now proceed entering in the simulation loop. At each iteration, the source vector is updated on the basis of the previous time step solution, with the freshly updated source vector a new nodal solution is computed by forward and backward substitution using the matrix L and U previously calculated and stored. The last step of the simulation loop, a part from saving the results, is to perform the components post step. With components post step we mean the computation of the quantities internal to each components that are needed to update the b vector at the next iteration. With reference to Example 2.5, we would have to calculate the voltage across and the current through the capacitor so that at the next time step the contribution of the capacitor to the source vector (3.31) can be calculated. The loop is repeated until the final simulation time imposed by the user is reached.

Example 2.5 Let us now consider the circuit of Figure 3.17 for which the parameters are reported in Table 3.2. Discretizing both the capacitor and the inductor with trapezoidal rule and applying the matrix stamp process the system in (3.32) is obtained.

$$\begin{bmatrix} \dfrac{1}{R_1}+\dfrac{2C}{h} & -\dfrac{2C}{h} \\[2ex] -\dfrac{2C}{h} & \dfrac{1}{R_2}+\dfrac{2C}{h}+\dfrac{h}{2L} \end{bmatrix} \begin{bmatrix} e_1(t+h) \\[1ex] e_2(t+h) \end{bmatrix}$$

$$= \begin{bmatrix} \dfrac{V}{R_1}+\dfrac{2C}{h}V_c(t)+I_c(t) \\[2ex] -\dfrac{h}{2L}V_L(t)-I_L(t)-\dfrac{2C}{h}V_c(t)-I_c(t) \end{bmatrix} \tag{3.32}$$

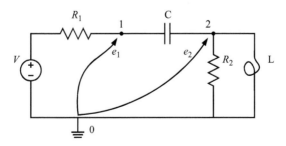

Figure 3.17   A simple example of linear dynamic circuit

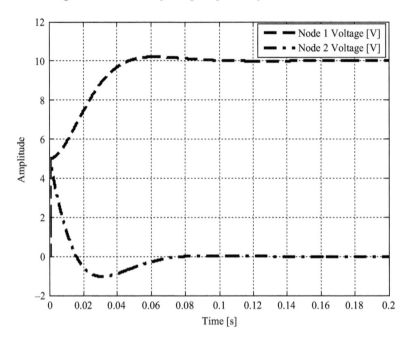

Figure 3.18   Example 2.5 solution

Applying the resistive companion solution flow of Figure 3.16, with a time step of 0.1 ms and a final time equal to 0.2 s, the results of Figure 3.18 are obtained.

3.4.2   Inductor and capacitor in resistive companion

In the previous section of this paragraph, we saw how to transform a capacitor in its DC equivalent using Euler backward, the purpose of this section is to show how we can discretize an inductor (see Figure 3.19) and a capacitor (see Figure 3.20) and representing it with the circuit of Figure 3.14, using Euler forward, Euler backward, and the trapezoidal rule.

3.4.2.1   Inductor

$$v_L = L\frac{di_L}{dt} \tag{3.33}$$

$$\frac{di_L}{dt} = \frac{v_L}{L} \tag{3.34}$$

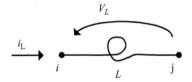

Figure 3.19   Inductor

3.4.2.2 Euler forward

Applying Euler forward integration method to the characteristic equation of the inductor, we obtain:

$$I_L(t+h) = I_L(t) + \frac{h}{L}V_L(t) \tag{3.35}$$

Since we want to map this equation to the circuit of Figure 3.14, we can write:

$$I_{Li}(t+h) = I_{Li}(t) + \frac{h}{L}V_L(t) \tag{3.36}$$

$$I_{Lj}(t+h) = -I_{Li}(t) - \frac{h}{L}V_L(t) \tag{3.37}$$

This means that the stamp for the inductor discretized with Euler forward is:

$$\begin{bmatrix} g & -g \\ -g & g \end{bmatrix} \begin{matrix} i \\ j \end{matrix} \tag{3.38}$$
$$\begin{matrix} i & j \end{matrix}$$

$$\begin{bmatrix} b(t) \\ -b(t) \end{bmatrix} \begin{matrix} i \\ j \end{matrix} \tag{3.39}$$

With:

$$g = 0 \tag{3.40}$$

$$b(t) = -I_L(t) - \frac{h}{L}V_L(t) \tag{3.41}$$

3.4.2.3 Euler backward

Applying Euler backward integration method to the characteristic equation of the inductor, we obtain:

$$I_L(t+h) = I_{Li}(t) + \frac{h}{L}V_L(t+h) \tag{3.42}$$

Since we want to map this equation to the circuit of Figure 3.14, we can write:

$$I_{Li}(t+h) = \frac{h}{L}V_L(t+h) + I_{Li}(t) \tag{3.43}$$

$$I_{Lj}(t+h) = -\frac{h}{L}V_L(t+h) - I_{Li}(t) \tag{3.44}$$

This means that the stamp for the inductor discretized with Euler backward is:

$$\begin{bmatrix} g & -g \\ -g & g \end{bmatrix} \begin{matrix} i \\ j \end{matrix} \tag{3.45}$$
$$\begin{matrix} i & j \end{matrix}$$

$$\begin{bmatrix} b(t) \\ -b(t) \end{bmatrix} \begin{matrix} i \\ j \end{matrix} \tag{3.46}$$

With:

$$g = \frac{h}{L} \tag{3.47}$$

$$b(t) = -I_{Li}(t) \tag{3.48}$$

3.4.2.4   Trapezoidal rule

Applying trapezoidal rule integration method to the characteristic equation of the inductor, we obtain:

$$I_L(t+h) = I_{Li}(t) + \frac{h}{2L}(V_L(t+h) + V_L(t)) \tag{3.49}$$

Since we want to map this equation to the circuit of Figure 3.14, we can write:

$$I_{Li}(t+h) = \frac{h}{2L}V_L(t+h) + I_{Li}(t) + \frac{h}{2L}V_L(t) \tag{3.50}$$

$$I_{Lj}(t+h) = -\frac{h}{2L}V_L(t+h) - I_{Li}(t) - \frac{h}{2L}V_L(t) \tag{3.51}$$

This means that the stamp for the inductor discretized with trapezoidal rule is:

$$\begin{bmatrix} g & -g \\ -g & g \end{bmatrix} \begin{matrix} i \\ j \end{matrix} \\ \begin{matrix} i & j \end{matrix} \tag{3.52}$$

$$\begin{bmatrix} b(t) \\ -b(t) \end{bmatrix} \begin{matrix} i \\ j \end{matrix} \tag{3.53}$$

With:

$$g = \frac{h}{2L} \tag{3.54}$$

$$b(t) = -I_{Li}(t) - \frac{h}{2L}V_L(t) \tag{3.55}$$

3.4.2.5   Capacitor

$$i_C = C\frac{dv_C}{dt} \tag{3.56}$$

$$\frac{dv_C}{dt} = \frac{i_C}{C} \tag{3.57}$$

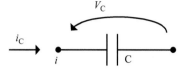

Figure 3.20   Capacitor

3.4.2.6   Euler forward

Applying Euler forward integration method to the characteristic equation of the capacitor, we obtain:

$$V_C(t+h) = V_C(t) + \frac{h}{C} I_C(t) \qquad (3.58)$$

As can been seen from previous equation, we miss the term $I_C(t+h)$ this means that we cannot map the capacitor discretized with Euler forward to the circuit of Figure 3.14 but the capacitor will be represented by an ideal voltage source with value equal to $V_C(t+h)$. This also implies that a circuit containing a capacitor discretized using Euler forward cannot be solved using nodal analysis but modified nodal analysis is required.

3.4.2.7   Euler backward

Applying Euler backward integration method to the characteristic equation of the capacitor, we obtain:

$$V_C(t+h) = V_C(t) + \frac{h}{C} I_C(t+h) \qquad (3.59)$$

Rearranging the previous equation, we obtain:

$$I_C(t+h) = \frac{C}{h} V_C(t+h) - \frac{C}{h} V_C(t) \qquad (3.60)$$

Since we want to map this equation to the circuit of Figure 3.14, we can write:

$$I_{Ci}(t+h) = \frac{C}{h} V_C(t+h) - \frac{C}{h} V_C(t) \qquad (3.61)$$

$$I_{Cj}(t+h) = -\frac{C}{h} V_C(t+h) + \frac{C}{h} V_C(t) \qquad (3.62)$$

This means that the stamp for the capacitor discretized with Euler backward is:

$$\begin{bmatrix} g & -g \\ -g & g \end{bmatrix} \begin{matrix} i \\ j \end{matrix} \qquad (3.63)$$
$$\begin{matrix} i & j \end{matrix}$$

$$\begin{bmatrix} b(t) \\ -b(t) \end{bmatrix} \begin{matrix} i \\ j \end{matrix} \qquad (3.64)$$

With:

$$g = \frac{C}{h} \tag{3.65}$$

$$b(t) = \frac{C}{h} V_C(t) \tag{3.66}$$

3.4.2.8   Trapezoidal rule

Applying trapezoidal rule integration method to the characteristic equation of the capacitor, we obtain:

$$V_C(t + h) = V_C(t) + \frac{h}{2C} (I_C(t + h) + I_C(t)) \tag{3.67}$$

Rearranging the previous equation, we obtain:

$$I_C(t + h) = \frac{2C}{h} V_C(t + h) - \frac{2C}{h} V_C(t) - I_C(t) \tag{3.68}$$

Since we want to map this equation to the circuit of Figure 3.14, we can write:

$$I_{Ci}(t + h) = \frac{2C}{h} V_C(t + h) - \frac{2C}{h} V_C(t) - I_{Ci}(t) \tag{3.69}$$

$$I_{Cj}(t + h) = -\frac{2C}{h} V_C(t + h) + \frac{2C}{h} V_C(t) + I_{Ci}(t) \tag{3.70}$$

This means that the stamp for the capacitor discretized with trapezoidal rule is:

$$\begin{bmatrix} g & -g \\ -g & g \end{bmatrix} \begin{matrix} i \\ j \end{matrix} \atop \begin{matrix} i & j \end{matrix} \tag{3.71}$$

$$\begin{bmatrix} b(t) \\ -b(t) \end{bmatrix} \begin{matrix} i \\ j \end{matrix} \tag{3.72}$$

With:

$$g = \frac{2C}{h} \tag{3.73}$$

$$b(t) = \frac{2C}{h} V_C(t) + I_{Ci}(t) \tag{3.74}$$

In Table 3.3, it is reported a summary of DC equivalence for capacitor and inductor.

Table 3.3 Summary of DC equivalence for capacitor and inductor

	Capacitor		Inductor	
	g	$b(t)$	g	$b(t)$
Euler forward	$V_C(t) + \frac{h}{C}I_C(t)$		0	$-I_L(t) - \frac{h}{L}V_L(t)$
Euler backward	$\frac{C}{h}$	$\frac{C}{h}V_C(t)$	$\frac{h}{L}$	$-I_{Li}(t)$
Trapezoidal rule	$\frac{2C}{h}$	$\frac{2C}{h}V_C(t) + I_{Ci}(t)$	$\frac{h}{2L}$	$-I_{Li}(t) - \frac{h}{2L}V_L(t)$

Example 2.6 We will now solve the same circuit of Example 2.5 showing step by step how Euler forward (EF), Euler backward (EB), and the trapezoidal rule (TR) can be used. Applying Euler forward to the circuit of Figure 3.17, we obtain the circuit of Figure 3.21.

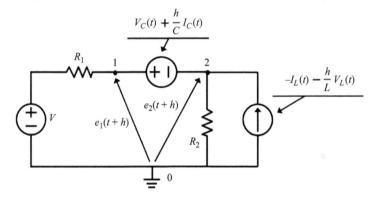

Figure 3.21 Euler forward integration of the circuit of Figure 3.17

Due to the presence of the ideal voltage source, we have to precede using modified nodal analysis. And so we obtain:

$$
\begin{bmatrix} \dfrac{1}{R_1} & 0 & 1 \\[2mm] 0 & \dfrac{1}{R_2} & -1 \\[2mm] 1 & -1 & 0 \end{bmatrix}
\begin{bmatrix} e_1(t+h) \\ e_1(t+h) \\ I_C(t+h) \end{bmatrix} =
\begin{bmatrix} \dfrac{V}{R_1} \\[2mm] -I_L(t) - \dfrac{h}{L}V_L(t) \\[2mm] V_C(t) + \dfrac{h}{C}I_C(t) \end{bmatrix}
\tag{3.75}
$$

One important topic, we just mentioned before is how to initialize the simulation. Every simulation algorithm must first be initialized. The initialization process depends on the specific algorithm adopted. Some algorithms may require for the initialization more information than what is typically provided as initial conditions.

As we said before, initial conditions are usually provided in the form of voltage across the capacitors and current through the inductors. Some integration methods such as e.g. EF require also the evaluation of the other component variable: current through the capacitor and voltage across the inductors. This problem can be solved by performing the solution of the circuit at $t=0^+$ where the initial conditions are forced and all the other variable are calculated. Notice that this solution is performed assuming that the time is frozen in a right neighborhood of $t=0$ forcing the initial conditions to be a continuous function, i.e.:

$$x(0^-) = x(0^+) \tag{3.76}$$

The circuit to be solved is reported in Figure 3.22. Notice that the circuit of Figure 3.22 does not depend from the specific integration method selected.

Assuming as initial conditions $V_{C0} = 0V$ and $I_{L0} = 0A$, we can calculate $I_C(0)$ and $V_L(0)$:

$$I_C(0) = 5A \tag{3.77}$$

$$V_L(0) = 5V \tag{3.78}$$

We can now proceed solving the transient using EF. Let us assume we decide to use $h = 0.1\text{ms}$

First of all, by using the two initial conditions just calculated, we can update the voltage source representing the capacitor and the current source representing the inductor:

$$V_C(1) = V_C(0) + \frac{h}{C}I_C(0) = 0.05 \text{ V} \tag{3.79}$$

$$I_L(1) = I_L(0) + \frac{h}{L}V_L(0) = 0.05 \text{ A} \tag{3.80}$$

Using the relation of (3.75), we can proceed calculating the node voltages and the current in the ideal source representing the capacitor (see circuit in Figure 3.22).

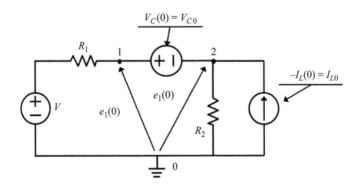

Figure 3.22   Initial conditions calculations

Those values should then be used to update the ideal source representing the capacitor and the ideal current source representing the inductor. This procedure is repeated over and over until the final simulation time is reached.

Let us now suppose we decide to solve the circuit by using EB. Applying Euler backward to the circuit of Figure 3.17, we obtain the circuit of Figure 3.23.

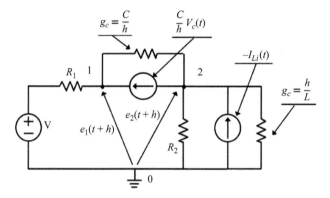

Figure 3.23   Euler backward integration of the circuit of Figure 3.17

The first main difference we can see is that we do not have any ideal voltage source and so we can precede using nodal analysis. And so we obtain:

$$\begin{bmatrix} \dfrac{1}{R_1}+\dfrac{C}{h} & -\dfrac{C}{h} \\ -\dfrac{C}{h} & \dfrac{1}{R_2}+\dfrac{C}{h}+\dfrac{h}{L} \end{bmatrix}\begin{bmatrix} e_1(t+h) \\ e_2(t+h) \end{bmatrix} = \begin{bmatrix} \dfrac{V}{R_1}+\dfrac{C}{h}V_c(t) \\ -I_L(t)-\dfrac{C}{h}V_c(t) \end{bmatrix}$$

(3.81)

The second one concerns the history component, since it only requires the value of the main state variable we do not need to pre-initialize the circuit solving for $t=0+$.

Let us now conclude this example looking at trapezoidal rule. Applying TR to the circuit of Figure 3.17, we obtain the circuit of Figure 3.24:

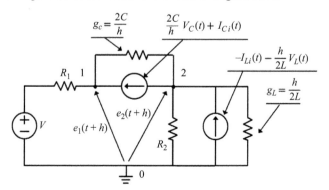

Figure 3.24   Trapezoidal rule integration of the circuit of Figure 3.17

As for Euler backward, since we do not have any ideal voltage source and so we can precede using nodal analysis. And so we obtain:

$$
\begin{bmatrix} \dfrac{1}{R_1} + \dfrac{2C}{h} & -\dfrac{2C}{h} \\[3mm] -\dfrac{2C}{h} & \dfrac{1}{R_2} + \dfrac{2C}{h} + \dfrac{h}{2L} \end{bmatrix} \begin{bmatrix} e_1(t+h) \\[2mm] e_2(t+h) \end{bmatrix}
$$

$$
= \begin{bmatrix} \dfrac{V}{R_1} + \dfrac{2C}{h} V_c(t) + I_c(t) \\[4mm] -\dfrac{h}{2L} V_L(t) - I_L(t) - \dfrac{2C}{h} V_c(t) - I_c(t) \end{bmatrix} \tag{3.82}
$$

In this case as for Euler forward, we also need to pre-initialize the circuit solving for $t=0+$. As mentioned before the circuit of Figure 3.22 should be used independently of the integration method used.

3.5 Numerical methods for the solution of linear systems

Until now we assume that the solution of the linear problem $Gx = b$ is obtained by left multiplying b with the inverse of G. This is normally not the case but numerical method can much more efficiently help us obtaining the solution (getting the value of x) without the need of invert matrix G. Among the all numerical methods developed to this purpose, two very popular are the Gaussian elimination method and the LU factorization. In this paragraph—purely as an example—we analyze the structure of the Gaussian elimination method.

3.5.1 Gaussian elimination

Let us suppose we want to solve the matrix problem:

$$
Ax = b \tag{3.83}
$$

where A as size $n \times n$ and x and b have size $n \times 1$.

The Gaussian elimination can be described through the following sequence of steps:

1. Create the augmented matrix "C."

$$
C = [A, b] \tag{3.84}
$$

2. Set an error flag to 1 ($E=1$).

$$
E = 1 \tag{3.85}
$$

3. For $j = 1 : n : \{$

 (a) Compute the pivot index

$$j \le i \le n \qquad (3.86)$$

 Such that:

$$\left|C_{p,j}\right| = \max\left(\left|C_{i,j}\right|\right) \qquad (3.87)$$

 (b) If $\left|C_{p,j}\right| = 0$, set the error flag to zero and exit

 (c) If $p \ge j$, interchange rows p and j

 (d) For each $i > j$, subtract $\frac{C_{ij}}{C_{ij}}$ times row j from row i

 $\}$

4. For $j = n : 1$ compute

$$x_j = \frac{1}{C_{j,j}}\left(C_{j,n+1} - \sum_{i=j+1}^{n} C_{j,i}x_i\right) \qquad (3.88)$$

Example 2.7 Let us consider the circuit in Figure 3.25.

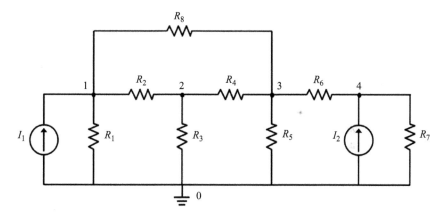

Figure 3.25   *Example circuit for Gaussian elimination*

Table 3.4   *Example 2.7 parameters*

R_1	R_8	$10\ \Omega$
R_2	$R_3\ R_7$	$5\ \Omega$
R_4	R_6	$2\ \Omega$
R_5		$1\ \Omega$
I_1		$1\ A$
I_2		$2\ A$

Let us build the conductance matrix by inspection using the values of Table 3.4:

$$
\begin{bmatrix}
\dfrac{1}{10}+\dfrac{1}{10}+\dfrac{1}{5} & -\dfrac{1}{5} & -\dfrac{1}{10} & 0 \\
-\dfrac{1}{5} & \dfrac{2}{5}+\dfrac{1}{2} & -\dfrac{1}{2} & 0 \\
-\dfrac{1}{10} & -\dfrac{1}{2} & 1+1+\dfrac{1}{10} & -\dfrac{1}{2} \\
0 & 0 & -\dfrac{1}{2} & \dfrac{1}{2}+\dfrac{1}{5}
\end{bmatrix}
\begin{bmatrix} v_1 \\ v_2 \\ v_3 \\ v_4 \end{bmatrix}
=
\begin{bmatrix} 1 \\ 0 \\ 0 \\ 2 \end{bmatrix}
\tag{3.89}
$$

$$
\begin{bmatrix}
0.4 & -0.2 & -0.1 & 0 \\
-0.2 & 0.9 & -0.5 & 0 \\
-0.1 & -0.5 & 2.1 & -0.5 \\
0 & 0 & -0.5 & 0.7
\end{bmatrix}
\begin{bmatrix} v_1 \\ v_2 \\ v_3 \\ v_4 \end{bmatrix}
=
\begin{bmatrix} 1 \\ 0 \\ 0 \\ 2 \end{bmatrix}
\tag{3.90}
$$

Row1–Row2

$$
\begin{bmatrix}
0.4 & -0.2 & -0.1 & 0 \\
0 & 0.8 & -0.55 & 0 \\
-0.1 & -0.5 & 2.1 & -0.5 \\
0 & 0 & -0.5 & 0.7
\end{bmatrix}
\begin{bmatrix} v_1 \\ v_2 \\ v_3 \\ v_4 \end{bmatrix}
=
\begin{bmatrix} 1 \\ 0.5 \\ 0 \\ 2 \end{bmatrix}
\tag{3.91}
$$

Row1–Row3

$$
\begin{bmatrix}
0.4 & -0.2 & -0.1 & 0 \\
0 & 0.8 & -0.55 & 0 \\
0 & -0.55 & 2.075 & -0.5 \\
0 & 0 & -0.5 & 0.7
\end{bmatrix}
\begin{bmatrix} v_1 \\ v_2 \\ v_3 \\ v_4 \end{bmatrix}
=
\begin{bmatrix} 1 \\ 0.5 \\ 0.25 \\ 2 \end{bmatrix}
\tag{3.92}
$$

Row1–Row4

$$
\begin{bmatrix}
0.4 & -0.2 & -0.1 & 0 \\
0 & 0.8 & -0.55 & 0 \\
0 & -0.55 & 2.075 & -0.5 \\
0 & 0 & -0.5 & 0.7
\end{bmatrix}
\begin{bmatrix} v_1 \\ v_2 \\ v_3 \\ v_4 \end{bmatrix}
=
\begin{bmatrix} 1 \\ 0.5 \\ 0.25 \\ 2 \end{bmatrix}
\tag{3.93}
$$

Row2–Row3

$$
\begin{bmatrix}
0.4 & -0.2 & -0.1 & 0 \\
0 & 0.8 & -0.55 & 0 \\
0 & 0 & 1.6969 & -0.5 \\
0 & 0 & -0.5 & 0.7
\end{bmatrix}
\begin{bmatrix} v_1 \\ v_2 \\ v_3 \\ v_4 \end{bmatrix}
=
\begin{bmatrix} 1 \\ 0.5 \\ 0.5938 \\ 2 \end{bmatrix}
\tag{3.94}
$$

Row2–Row4

$$
\begin{bmatrix}
0.4 & -0.2 & -0.1 & 0 \\
0 & 0.8 & -0.55 & 0 \\
0 & 0 & 1.6969 & -0.5 \\
0 & 0 & -0.5 & 0.7
\end{bmatrix}
\begin{bmatrix} v_1 \\ v_2 \\ v_3 \\ v_4 \end{bmatrix}
=
\begin{bmatrix} 1 \\ 0.5 \\ 0.5938 \\ 2 \end{bmatrix}
\tag{3.95}
$$

Row3–Row4

$$\begin{bmatrix} 0.4 & -0.2 & -0.1 & 0 \\ 0 & 0.8 & -0.55 & 0 \\ 0 & 0 & 1.6969 & -0.5 \\ 0 & 0 & 0 & 0.5527 \end{bmatrix} \begin{bmatrix} v_1 \\ v_2 \\ v_3 \\ v_4 \end{bmatrix} = \begin{bmatrix} 1 \\ 0.5 \\ 0.5938 \\ 2.1750 \end{bmatrix} \qquad (3.96)$$

Now we can solve by back substitution:

$$v_4 = \frac{2.1750}{0.5527} = 3.9354 \qquad (3.97)$$

$$v_3 = \frac{0.5v_4 + 0.5938}{1.6969} = 1.5095 \qquad (3.98)$$

$$v_2 = \frac{0.55v_3 + 0.5}{0.8} = 1.6682 \qquad (3.99)$$

$$v_1 = \frac{0.2v_2 + 0.1v_3 + 1}{0.4} = 3.7088 \qquad (3.100)$$

3.6   Controlled sources

An important family of components for modeling of electrical circuit is the one of the controlled sources. Controlled sources can be used to model a variety of components (e.g. small signal representation of transistors, operational amplifiers, ideal transformers). Controlled source can be of four different categories:

- Voltage controlled voltage source (VCVS)
- Voltage controlled current source (VCCS)
- Current controlled current source (CCCS)
- Current controlled voltage source (CCVS)

Among the four different types only the VCCS can be represented in classical nodal analysis. For the three other categories, we need to refer to modified nodal analysis or insert suitable resistances to transform each of the other three cases in a VCCS.

3.6.1   VCCS

This is the only type that can rigorously be represented in classical nodal analysis. Like any other controlled source it is a four terminal device:

- Two terminals are used for the controlling variable.
- Two terminals are used for the controlled variable.

In this case, we have (see Figure 3.26):

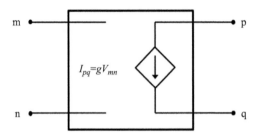

Figure 3.26   VCCS

The characteristic equation as reported in the picture is:

$$I_{pq} = gV_{mn} \qquad (3.101)$$

where g has the units of a conductance but it is not physically speaking a resistance. It is usually referred as trans-conductance. We can define a matrix stamp process for this device as reported in the following:

	m	n	p	q
m				
n				
p	g	-g		
q	-g	g		

$$(3.102)$$

Notice that it looks like the matrix stamp of a resistance with conductance g but the important difference is that the rows and columns are not with the same indices. As result of that, the symmetry of the conductance matrix is lost.

3.6.2   VCVS

For this type of device (see Figure 3.27), the characteristic equation is:

$$V_{pq} = \alpha V_{mn} \qquad (3.103)$$

This equation because does not give explicitly the current through the terminals p and q, as consequence classical nodal analysis, cannot be used.

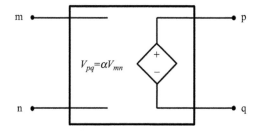

Figure 3.27   VCVS

It is possible to extend modified nodal analysis to include this type of device by using the same approach adopted for a standard ideal voltage source. We introduce as new unknown the current through the terminals p and q and then we write an extra equation.

Following this approach, the matrix stamp will look like:

	m	n	p	q	t
m					
n					
p					1
q					-1
t	$-\alpha$	α	1	-1	

$$(3.104)$$

t is the additional line included due to the ideal voltage source between p and q.

3.6.3   CCCS

For this type of device (see Figure 3.28), the characteristic equation is:

$$I_{pq} = \beta I_{mn} \tag{3.105}$$

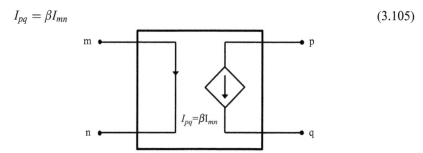

Figure 3.28   CCCS

Also in this case, the source cannot be immediately represented in standard nodal analysis because the current is not available as explicit variable. We can treat the shortcut between m and n as an ideal voltage source with value equal to zero. This will require again the use of modified nodal analysis.

Following this approach, the matrix stamp will look like:

	m	n	p	q	z
m					1
n					−1
p					β
q					$-\beta$
z	1	−1			

$$(3.106)$$

z is the additional line included due to the ideal voltage source between *m* and *n*.

3.6.4   CCVS

For this type of device (see Figure 3.29), the characteristic equation is:

$$V_{pq} = rI_{mn} \qquad (3.107)$$

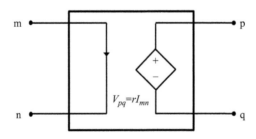

$$V_{pq} = rI_{mn}$$

Figure 3.29   CCVS

Also in this case, we need to adopt a modified nodal analysis approach to properly represent the controlled source. Because we have two problems to solve the current as an input variable and the voltage as an output variable, we need to introduce two new unknowns and two new equations.

Following this approach, the matrix stamp will look like:

	m	n	p	q	z	t
m					1	
n					−1	
p						1
q						−1
z	1	−1				
t			1	−1	−r	

$$(3.108)$$

z is the additional line included due to the ideal voltage source between m and n, t is the additional line included due to the ideal voltage source between p and q.

3.6.4.1 Non-linear resistive companion

In the previous paragraphs, we have seen how to solve linear dynamic circuits using nodal analysis and resistive companion, but the big family of the non-linear components has not been considered yet. In this chapter, we will extend nodal analysis and resistive companion to be able to treat non-linear components when cannot be reduced to the solution of piecewise linear circuit [7].

Let us first start taking a look to the solution of a simple non-linear algebraic equation [8], like the one in (3.109)

$$x = 4 - 2x^{\frac{1}{3}} \triangleq F(x) \tag{3.109}$$

Our objective is to define a method to find $x = x^*$ where x^* is a value that reduces (3.109) to an identity. The solution to this problem can be obtained through the use of the so-called fixed-point iteration algorithm as described by the flow chart in Figure 3.30.

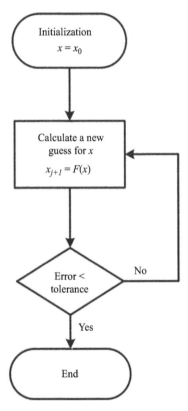

Figure 3.30   Fixed-point iteration algorithm flow chart

According to the fixed point algorithm starting from an initial guess of $x = x_0$, we compute $F(x_0) \triangleq x_1$, the process is iterated until $|x_j - x_{j+1}| < \varepsilon$, where ε is the error tolerance. This algorithm relays on the fact that, under certain conditions, $|x^* - x_1| < |x^* - x_0|$.

In Table 3.5, it is reported the solution of (3.109) when the initial condition is equal to -10 and $\varepsilon = 0.0001$.

Table 3.5   *Fixed point iteration for (3.109)*

j	0	1	2	3	4	5
x_j	6	0.3658	2.5697	1.2606	1.8395	1.5495

j	6	7	8	9	10	11
x_j	1.6857	1.6198	1.6512	1.6361	1.6433	1.6399

j	12	13	14	15	16	17
x_j	1.6415	1.6407	1.6411	1.6409	1.6410	1.6410

Generalizing the previous example, we can say that assuming to have a system of non-linear equations in the form of $x = F(x)$, the fixed point algorithm can be summarized by (3.110):

$$x_{j+1} = F(x_j) \tag{3.110}$$

Under appropriate conditions, this sequence converges to a fixed point that is the solution of the system of (3.111):

$$\lim_{j=\infty} x_j = x^* \tag{3.111}$$

A simple criterion that guarantees convergence is given by the principle of contraction mapping.

If $F(x)$ is a contraction from the n-dimension space R^n into R^n and if there exists a constant $L<1$ such that:

$$\|F(x) - F(y)\| \leq L\|x - y\| \tag{3.112}$$

for all $x \in R^n$ and $y \in R^n$, then $F(x)$ has a unique fixed point $x = x^*$ and the sequence in (3.110) converges to this fixed point. An upper bound on the error for stopping at the jth iteration is given by:

$$\|x^* - x_j\| \leq \frac{L}{1-L}\|x_1 - x_0\|$$

Applying what has been just presented to resistive companion the method can be extended to the solution of non-linear network [1,9]. Let us now look to the nodal equation of (3.6). It can be transformed in the fixed point form $x = F(x)$.

We consider two type of non-linear components: a voltage controlled non-linear resistor (3.113) and a VCCS (3.114)

$$i_k = g(v_k) \tag{3.113}$$

$$i_k = g(v_j) \tag{3.114}$$

Considering the generic branch of Figure 3.31, we can reformulate nodal equations, also including non-linear components.

$$\widehat{A}i = 0 \tag{3.115}$$

where:

$$\widehat{i} = \begin{bmatrix} \widehat{i}_1 & \widehat{i}_2 & \dots & \widehat{i}_b \end{bmatrix}^T \tag{3.116}$$

But we can also write:

$$\widehat{i} = i - J \tag{3.117}$$

$$i = \begin{bmatrix} i_1 & i_2 & \dots & i_b \end{bmatrix}^T \tag{3.118}$$

$$J = \begin{bmatrix} J_1 & J_2 & \dots & J_b \end{bmatrix}^T \tag{3.119}$$

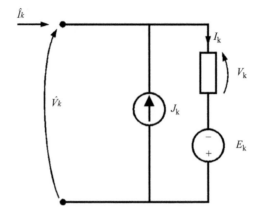

Figure 3.31   Generic branch

And for the non-linear components:

$$i = \begin{bmatrix} i_1 \\ i_2 \\ \vdots \\ i_b \end{bmatrix} = \begin{bmatrix} g_1(v_\alpha) \\ g_2(v_\beta) \\ \vdots \\ g_b(v_\gamma) \end{bmatrix} = g(v) \tag{3.120}$$

So we obtain:

$$Ag(v) = AJ \tag{3.121}$$

Considering that:

$$\hat{v} = v - E \tag{3.122}$$

$$\hat{v} = A^t v_n \tag{3.123}$$

We obtain:

$$Ag(\hat{v} + E) = AJ \tag{3.124}$$

$$Ag(A^t v_n + E) = AJ \tag{3.125}$$

For an $n+1$ node network, this represents a system of n non-linear equation in the node voltages

Since our goal is to apply the fix point algorithm to the solution of equation of (3.125) we have to manipulate a little this equation. Obtaining:

$$Ag(A^t v_n + E) - AJ = 0 \tag{3.126}$$

And defining:

$$F(x) = x - K(x)f(x) \tag{3.127}$$

Note the we can now apply fix point algorithm and that any solution of $x = x^*$ of $f(x)$ is also a fixed point of $F(x)$. $K(x)$ is an (n,n) matrix function of x, non-singular for any $f(x) = 0$. Depending on how $K(x)$ is defined different rate of convergence can be obtained; therefore different iteration technique can be defined according to the specific $K(x)$. A very common and well known approach is the Newton–Raphson, in this case the $K(x)$ is defined as the inverse of the Jacobian

$$K(x) = J^{-1}(x) \tag{3.128}$$

where:

$$J(x) \triangleq \begin{bmatrix} \dfrac{\partial f_1(x)}{\partial x_1} & \dfrac{\partial f_1(x)}{\partial x_2} & \cdots & \cdots & \dfrac{\partial f_1(x)}{\partial x_n} \\[2mm] \dfrac{\partial f_2(x)}{\partial x_1} & \dfrac{\partial f_2(x)}{\partial x_2} & \cdots & \cdots & \dfrac{\partial f_2(x)}{\partial x_n} \\[2mm] \vdots & & & & \vdots \\[2mm] \vdots & & & & \vdots \\[2mm] \dfrac{\partial f_n(x)}{\partial x_1} & \dfrac{\partial f_n(x)}{\partial x_2} & \cdots & \cdots & \dfrac{\partial f_n(x)}{\partial x_n} \end{bmatrix} \tag{3.129}$$

And so the solution of the non-linear problem becomes:

$$x_{j+1} = x_j - \left[J(x_j) \right]^{-1} f(x_j) \tag{3.130}$$

If now we proceed applying the Newton–Raphson method to the nodal equations, we obtain:

$$f(v_n) = Ag(A^t v_n + E) - AJ \tag{3.131}$$

$$J(v_n) = A \frac{\partial g(A^t v_n + E)}{\partial v} A^t \tag{3.132}$$

$$v_n^{j+1} = v_n^j - \left[A \frac{\partial g(A^t v_n^j + E)}{\partial v} A^t \right]^{-1} \left[Ag(A^t v_n^j + E) - AJ \right] \tag{3.133}$$

If now we define:

$$E_Q^j = A^t v_n^j + E \tag{3.134}$$

$$J_Q^j = g(A^t v_n^j + E) = g\left(E_Q^j \right) \tag{3.135}$$

$$G_Q^j = \frac{\partial g(A^t v_n^j + E)}{\partial v} = \frac{\partial g\left(E_Q^j \right)}{\partial v} \tag{3.136}$$

We can write:

$$v_n^{j+1} = v_n^j - \left[A G_Q^j A^t \right]^{-1} \left[A J_Q^j - AJ \right] \tag{3.137}$$

$$\left[AG_Q^j A^t \right] v_n^{j+1} = A \left[J - J_Q^j + G_Q^j A^t v_n^j \right] \tag{3.138}$$

$$\left[AG_Q^j A^t \right] v_n^{j+1} = A \left[J - J_Q^j + G_Q^j \left(E_Q^j - E \right) \right] \tag{3.139}$$

Equation (3.139) can be mapped to a DC equivalent circuit to be solved at each iteration. If for sake of simplicity, we consider the E and J equal to zero. We can represent the non-linear resistor of Figure 3.32(a) with the equivalent circuit of Figure 3.32(b).

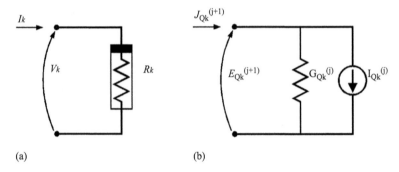

(a)   (b)

Figure 3.32   *(a) Non-linear resistor and (b) non-linear resistor model at the jth iteration*

Where:

$$G_{Qk}^{(j)} = \left. \frac{dg_k(v_k)}{dv_k} \right|_{E_{Qk}^{(j)}} \tag{3.140}$$

$$I_{Qk}^{(j)} = J_{Qk}^{(j)} - G_{Qk}^{(j)} E_{Qk}^{(j)} \tag{3.141}$$

$J_{Qk}^{(j)}$ is the current through resistor R_k at the jth iteration; $E_{Qk}^{(j)}$ is the voltage across resistor R_k at the jth iteration.

3.6.4.2   Non-linear resistive companion solution flow

Let us now look at how the solution flow of resistive companion, previously shown in Figure 3.16, gets modified to include the solution of non-linear components. Comparing Figure 3.33 with Figure 3.16, the first difference to notice is that now in addition to the external loop regulated by the imposed timeline there is another loop, internal to the first one. This loop is needed to allow the solution of non-linear components according with the method presented. All non-linear components contributions will have to be updated at each iteration (with reference to the non-linear resistor model presented before, (3.140) and (3.141)) and the nodal solution repeated until a predefined tolerance is reached. Updating the conductance values has another important impact on the execution flow: the conductance matrix G is

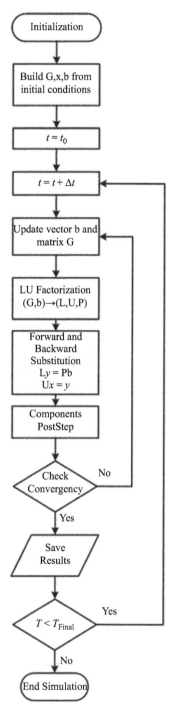

Figure 3.33 Non-linear resistive companion solution flow

not anymore constant during the simulation: as consequence its factorization cannot be performed off-line but has to be repeated even multiple times during each time step. Looking at the two loops depicted in Figure 3.33 another important consideration has to be done. While the number of iterations in the first loop can be easily and exactly predicted, knowing the final simulation time and the step size, the number of iterations in the second loop depends on the selected accuracy as well as on the specific operating point and cannot be defined a priori. The two considerations just presented strongly limit the applicability of the described solution scheme to real time simulation.

Exercises

Exercise 1 Let us consider the circuit of Figure 3.34 and the parameter in Table 3.6.

(A)   Build conductance matrix (G) and source vector (b) by applying nodal analysis method.
(B)   Determine the voltage value of each node.

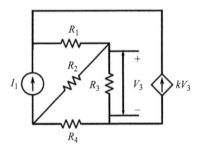

Figure 3.34   Circuit for Exercise 1

Table 3.6   Exercise 1 parameters

I_1	10 A
R_1, R_2	1 Ω
R_3, R_4	3 Ω
k	2 A/V

Exercise 2 Given the circuit of Figure 3.34 build the conductance matrix (G) and source vector (b) by applying modified nodal analysis method.

Exercise 3 Given the circuit of Figure 3.35 build the conductance matrix (G) and source vector (b) by applying modified nodal analysis method.

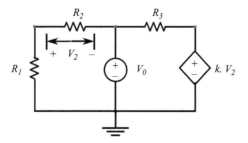

Figure 3.35   Circuit for Exercise 2

Exercise 4 Given the circuit of Figure 3.36 build the conductance matrix (G) and source vector (b) by applying modified nodal analysis method.

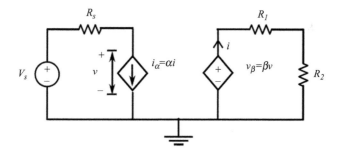

Figure 3.36   Circuit for Exercise 3

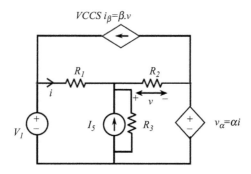

Figure 3.37   Circuit for Exercise 4

Exercise 5 Given the circuit of Figure 3.38 and the parameters from Table 3.7 solve it for two time steps using resistive companion method and trapezoidal rule for discretization.

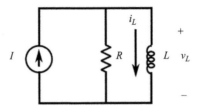

Figure 3.38   Circuit for Exercise 5

Table 3.7   Exercise 5 parameters

I	10 A
R	0.2 Ω
L	1 mH
t	10 ms
$I_L(0)$	0 A

Exercise 6 Given the circuit of Figure 3.39 and the parameters in Table 3.8 solve it for two time steps using resistive companion method and Euler backward for discretization.

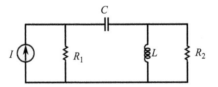

Figure 3.39   Circuit for Exercise 6

Table 3.8   Exercise 6 parameters

I	10 A
R_1, R_2	1 Ω
L	1 mH
t	1 ms
$I_L(0)$	0 A
$V_C(0)$	0 V

Exercise 7 Given the circuit of Figure 3.40 and the parameters in Table 3.9 solve it for two time steps using resistive companion method and Euler backward for discretization and the parameters as from Table 3.9.

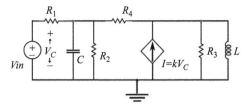

Figure 3.40   Circuit for Exercise 7

Table 3.9   Exercise 7 parameters

V_{in}	10 V
R_1, R_2, R_3, R_4	1 Ω
C	1 mF
L	1 mH
k	1 A/V
t	1 ms
$I_L(0)$	0 A
$V_C(0)$	0 V

Exercise 8 Consider the circuit of Figure 3.41 and the parameter in Table 3.10, derive the admittance and source matrix of this circuit and solve the circuit for two time step. Notice that the characteristic equation of the non-linear resistor—reported in (3.142)—requires the use of non-linear resistive companion method.

$$i_R = g(v_R) = \frac{v_R^2}{2} \tag{3.142}$$

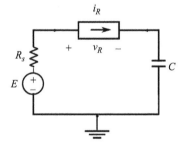

Figure 3.41   Circuit for Exercise 8

Table 3.10   Exercise 8 parameters

E	10 V
R_s	1 Ω
C	10 µF
t	1 µs
$V_C(0)$	0 V

References

[1] L.O. Chua, P.-M. Lin, *"Computer-Aided Analysis of Electronic Circuits: Algorithms and Computational Techniques"*, Prentice-Hall Inc., Englewood Cliffs, NJ, 1975.

[2] H.W. Dommel, "Digital computer solution of electromagnetic transients in single and multiphase networks", *IEEE Transactions on Power Apparatus and Systems,* 1969, Vol. PAS-88, no. 4, 388–399.

[3] P. Dimo, "Nodal analysis of power systems", Gordon & Breach Science Pub, Philadelphia, PA, 1975, ISBN 0-85626-001-0.

[4] C.W. Ho, A.E. Ruehli, P.A. Brennan, "The modified nodal approach to network analysis", *IEEE Transactions on Circuits and Systems,* 1975, Vol. CAS-22, 504–509.

[5] H.W. Dommel, "Computation of electromagnetic transients", *Proceedings of the IEEE,* 1974, Vol. 62, no. 7, 983–993.

[6] H.W. Dommel, "Fast transient stability solutions", *IEEE Transactions on Power Apparatus and Systems,* 1972, Vol. PAS-91, no. 4, 1643–1650.

[7] L. Chua, "Efficient computer algorithms for piecewise-linear analysis of resistive nonlinear networks", *IEEE Transactions on Circuit Theory,* 1971, Vol. 18, no. 1, 73–85.

[8] J.M. Ortega, W.C. Rheinboldt, *"Iterative Solution of Nonlinear Equations in Several Variables"*, Academic Press, New York, NY, 1970.

[9] H.W. Dommel, "Nonlinear and time-varying elements in digital simulation of electromagnetic transients", *IEEE Transactions on Power Apparatus and Systems*, 1971, Vol. PAS-90, no. 6, 2561–2256.

Chapter 4

State-space methods

Andrea Benigni[1,2] and Antonello Monti[3,4]

Besides nodal methods another very common and widely used family of modeling approaches is the one based on state-space representation. Simulations solvers and tools based on this approach typically implement what is normally referred as signal flow solution process, the most well known commercial tool of this type is Simulink® from MathWorks®. This type of solver has been widely used for the modeling of mechanical, thermal, and hydraulic systems but it was less successful in the modeling of electrical systems. First of all, explicit integration better fits signal flow solvers but at the same time explicit integration methods often are not suitable for the integration of electrical systems, especially if the interest is in using relatively large time steps. Moreover, those types of solvers—as state-space modeling approach in general—make the automatic creation of system level models by coupling of components models difficult and not always possible; this is extremely critical for large power systems composed of hundreds of components. At the same time, state-space modeling is also extremely important as modeling method for control engineering independently of its use for simulation purposes. In this chapter, we provide a concise review of state-space modeling and its use for the simulation of electrical systems, for the reader interested in diving more in deep on the topic [1] represent a good starting point for the use of state-space modeling in simulation and [2] for the use of state-space modeling for control purposes.

The state-space description of a system is based on a set of coupled first-order differential equations in a set of internal variables named state variables, and a set of algebraic equations that on the based of state variable and system inputs define a set of output variables.

Let us start providing a definition for the "state of a system." In a static system, the outputs depend only on the present values of its inputs, vice versa in a dynamic system the outputs depend on the present and past values of its inputs. The state of a dynamic system contains all the information needed to calculate

[1]Institute of Energy and Climate Research: Energy Systems Engineering (IEK-10), Juelich Research Center, Germany
[2]Department of Mechanical Engineering, RWTH Aachen University, Germany
[3]Institute for Automation of Complex Power Systems, RWTH Aachen University, Germany
[4]Fraunhofer FIT Center for Digital Energy, Germany

responses to present and future inputs without reference to the past history of its inputs and outputs.

The state of a system at any time t_c is the minimal amount of information that, together with all inputs for $t > t_c$, uniquely determines the behavior of the system for all $t > t_c$.

This definition asserts that the dynamic behavior of a state-determined system is completely characterized by the response of the set of n state variables, where n is the order of the system. It should be observed that it does not exist a unique set of state variables describing any given system but many different sets of state variables may be defined. Vice versa, for a given system the order n is unique, and is independent of the particular set of state variables chosen. State variable systems descriptions may be formulated in terms of physical and measurable variables, or in terms of variables that are not directly measurable and it is also possible to mathematically transform one set of state variables to another. The important point is that any set of state variables must provide a complete description of the system. For a linear system the number of state variable is equal to the number of independent energy storage elements in the system. In many practical cases, the variables representing the energy stored in each storage element are selected to be state variables, in this way, the total system energy can be directly calculated and the time derivatives of the state variables determine the rate of change of the system energy. While this is a very common and convenient choice, the only requirement selecting a state variable is that it should be continuous and differentiable: a state variable does not necessary need to be associated with an energy storage element.

Considering the definition of "state of a system" state-space methods should be preferred to topological (node or mesh) ones due to the minor number of variables to be calculated. In fact, in an electrical circuit, the number of nodes or meshes often is significantly larger than the number of state variables. In this context, it is clear that state-space-based simulation requires much less calculation than nodal analysis. At the same time, the computer implementation of state-space models is more complicated that the one of nodal ones: it often requires the direct involvement of a user with a significant experience in modeling, it is extremely error prone and it may lead to the creation of simulation artifacts hard to identify. Moreover, if implicit integration is required by the nature of the system, and this is often the case for electrical systems, part of the advantage of state-space methods is lost.

We now proceed describing the basic principles of state-space modeling and how a state-space model of an electric circuit can be obtained and solved, we will also introduce a method that provide a way to automatically generate state-space models from system topology and predefined components structure. To conclude the chapter the relation between state-space and transfer function representations of a circuit is presented.

4.1   State-space modeling

The mathematical description of a system based on state-space modeling is obtained by a set of n coupled first-order ordinary differential equations—known as the state equations—in which the time derivative of each state variable $\dot{x}_i(t)$ is expressed in terms of the state variables $x_1(t), x_2(t), \ldots\ldots, x_n(t)$ and of the inputs $u_1(t), u_1(t), \ldots\ldots, u_m(t)$ of the system, where m is the number of inputs of the system. Assuming the general case in which the relation between the derivatives of each state variable and the state variables and the inputs is nonlinear we have:

$$\dot{x}_1(t) = f_1(x_1(t), x_2(t), \ldots\ldots, x_n(t), u_1(t), u_2(t), \ldots\ldots, u_m(t), t)$$
$$\dot{x}_2(t) = f_2(x_1(t), x_2(t), \ldots\ldots, x_n(t), u_1(t), u_2(t), \ldots\ldots, u_m(t), t)$$

$$\vdots = \vdots$$

$$\dot{x}_n(t) = f_n(x_1(t), x_2(t), \ldots\ldots, x_n(t), u_1(t), u_2(t), \quad\ldots\ldots, u_m(t), t) \tag{4.1}$$

Defining the state vector $\mathbf{x}(t) \in \mathbb{R}^n$ and the input vector $\mathbf{u}(t) \in \mathbb{R}^m$, we can write the state model of a generic non-linear system as:

$$\dot{x}(t) = f(x(t), u(t), t) \tag{4.2}$$

If now we focus our attention on linear and time-invariant systems (LTI), we can express the relation between each state variable derivative $\dot{x}_i(t)$ as the weighted sum of the state variables $x_1(t), x_2(t), \ldots\ldots, x_n(t)$ and the system inputs $u_1(t), u_1(t), \quad\ldots\ldots, u_m(t)$.

$$\dot{x}_1(t) = a_{11}x_1(t) + a_{12}x_2(t) + \ldots + a_{1n}x_n(t) + b_{11}u_1(t) + b_{12}u_2(t)$$
$$+ \ldots b_{1m}u_m(t)$$

$$\dot{x}_2(t) = a_{21}x_1(t) + a_{22}x_2(t) + \ldots + a_{2n}x_n(t) + b_{21}u_1(t) + b_{22}u_2(t)$$
$$+ \ldots b_{2m}u_m(t)$$

$$\vdots = \vdots$$

$$\dot{x}_n(t) = a_{n1}x_1(t) + a_{n2}x_2(t) + \ldots + a_{nn}x_n(t) + b_{n1}u_1(t) + b_{n2}u_2(t)$$
$$+ \ldots b_{nm}u_m(t) \tag{4.3}$$

Writing (4.3) in a matrix form using the state vector $x(t)$ and the input vector $u(t)$, we obtain:

$$\dot{x}(t) = Ax(t) + Bu(t) \tag{4.4}$$

where $A \in \mathbb{R}^{n \times n}$ is usually referred to as the system matrix and $B \in \mathbb{R}^{n \times m}$ as the input matrix:

$$A = \begin{bmatrix} a_{11} & a_{12} & \cdots & a_{1n} \\ a_{21} & \ddots & & \\ \vdots & & \ddots & \\ a_{n1} & & & a_{nn} \end{bmatrix} \tag{4.5}$$

The system matrix (4.5) relates how the current state $x_i(t)$ affects the state change $\dot{x}_j(t)$ by a constant coefficient a_{ji}:

$$B = \begin{bmatrix} b_{11} & b_{12} & \cdots & b_{1m} \\ b_{21} & \ddots & & \\ \vdots & & \ddots & \\ b_{n1} & & & b_{nm} \end{bmatrix} \tag{4.6}$$

The input matrix (4.6) determines how the system inputs $u_i(t)$ affect the state change $\dot{x}_j(t)$ by a constant coefficient b_{ji}.

In many practical cases, the description of a physical system in terms of a set of state variables does not necessarily include all of the variables of engineering interest. In this context, it is important to define a set of output equations that relate each output variable $y_i(t)$ to the state variables $x_1(t), x_2(t), \ldots\ldots, x_n(t)$ and the inputs $u_1(t), u_1(t), \ldots\ldots, u_m(t)$ of the system. Assuming the general case in which the relation between the output variables, the state variables and the inputs are nonlinear we have:

$$y_1(t) = g_1(x_1(t), x_2(t), \ldots\ldots, x_n(t), u_1(t), u_2(t), \ldots\ldots, u_m(t), t)$$

$$y_2(t) = g_2(x_1(t), x_2(t), \ldots\ldots, x_n(t), u_1(t), u_2(t), \ldots\ldots, u_m(t), t)$$

$$\vdots = \vdots$$

$$y_p(t) = g_p(x_1(t), x_2(t), \ldots\ldots, x_n(t), u_1(t), u_2(t), \ldots\ldots, u_m(t), t) \tag{4.7}$$

where p is the number of output variables, defining the output vector $y(t) \in \mathbb{R}^p$ we can write the output equation for a generic non-linear system as:

$$y(t) = g(x(t), u(t), t) \tag{4.8}$$

If, as before, we focus our attention on linear and time invariant systems (LTI), we can express the relation between each output variable $y_i(t)$ as the weighted sum of the state variables $x_1(t), x_2(t), \ldots\ldots, x_n(t)$ and the system inputs $u_1(t), u_1(t), \ldots\ldots, u_m(t)$:

$$y_1(t) = c_{11}x_1(t) + c_{12}x_2(t) + \ldots + c_{1n}x_n(t) + d_{11}u_1(t) + d_{12}u_2(t)$$

$$+ \ldots d_{1m}u_m(t)$$

$$y_2(t) = c_{21}x_1(t) + c_{22}x_2(t) + \ldots + c_{2n}x_n(t) + d_{21}u_1(t) + d_{22}u_2(t)$$

$$+ \ldots d_{2m}u_m(t)$$

$$\vdots = \vdots$$

$$y_p(t) = c_{p1}x_1(t) + c_{p2}x_2(t) + \ldots + c_{pn}x_n(t) + d_{p1}u_1(t) + d_{p2}u_2(t)$$

$$+ \ldots d_{pm}u_m(t) \tag{4.9}$$

Writing (4.9) in a matrix form using the output $y(t)$, we obtain:

$$y(t) = Cx(t) + Du(t) \tag{4.10}$$

where $C \in \mathbb{R}^{p \times n}$ is usually referred to as the output matrix and $D \in \mathbb{R}^{p \times m}$ as the feed-forward matrix:

$$C = \begin{bmatrix} c_{11} & c_{12} & \cdots & c_{1n} \\ c_{21} & \ddots & & \\ \vdots & & \ddots & \\ c_{p1} & & & c_{pn} \end{bmatrix} \tag{4.11}$$

The output matrix (4.11) relates how the current state $x_i(t)$ affects the system outputs $y_i(t)$ by a constant coefficient c_{ji}:

$$D = \begin{bmatrix} d_{11} & d_{12} & \cdots & d_{1m} \\ d_{21} & \ddots & & \\ \vdots & & \ddots & \\ d_{p1} & & & d_{pm} \end{bmatrix} \tag{4.12}$$

The feed-forward matrix (4.12) determines how the system inputs $u_i(t)$ affect the system outputs $y_i(t)$ by a constant coefficient d_{ji}.

Summarizing, the state-space model of a linear and time invariant systems (LTI) is described by two equations:

$$\dot{x}(t) = Ax(t) + Bu(t)$$
$$y(t) = Cx(t) + Du(t) \tag{4.13}$$

4.2   Circuit modeling

As mentioned before, the creation of the state-space model of an electrical circuit is not always a trivial procedure and experience plays a significant role. In any case, even if, unlike the resistive companion, a step by step fully formalized procedure is hard to define a good procedure to create a state-space model of an electrical circuit is illustrated by the following steps:

1. Identify the number of independent variables n.
2. Select a set of state variables $x(t)$.
3. Write a mesh equation using KVL for each independent inductor.
4. Write a node equation using KCL for each independent capacitor.
5. Verify that the sum of KCL and KVL equations is equal to n. It's important to remember that the same KCL and KVL can never be used two times.
6. Using the needed algebraic equations re-arrange the n equations so that the only terms involved are the state variables and their derivatives $(x(t), \dot{x}(t))$, the circuit inputs $u(t)$ and the circuit parameters.

7.   Organize the *n* equations in the form $\dot{x}(t) = Ax(t) + Bu(t)$.
8.   Identify the needed outputs of interest in number equal to *p*.
9.   Write *p* algebraic equations, one for each output. Each equation should include only one output, the state variables vector $x(t)$, the circuit inputs vector $u(t)$, and the circuit parameters.
10. Organize the *p* equations in the form $y(t) = Cx(t) + Du(t)$.

Example 4.1 To better illustrate this procedure, let us consider the circuit in Figure 4.1.

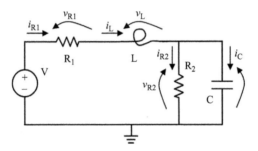

Figure 4.1   Linear dynamic circuit

1.   In the circuit, there are two independent energy storage elements, one capacitor and one inductor: $n = 2$.
2.   As state variables, we can select the current through the inductor $i_L(t)$ and the voltage across the capacitor $v_C(t)$.
3.   We proceed writing a mesh equation using KVL for the inductor:

$$v_L(t) + v_{R1}(t) - V + v_C(t) = 0 \qquad (4.14)$$

4.   We write a node equation using KCL for the capacitor:

$$i_L(t) - i_C(t) - i_{R2}(t) = 0 \qquad (4.15)$$

5.   We have one KCL and one KVL so we have a total of two equations that equal the number of state variables *n*.
6.   Using the needed algebraic equations, we re-arrange the *n* equations so that the only terms involved are the state variables and their derivatives $\left(i_L(t),\ v_C(t), \frac{di_L(t)}{dt}, \frac{dv_C(t)}{dt}\right)$, the circuit inputs *V* and the circuit parameters (R_1, R_2, L, C):

$$L\frac{di_L(t)}{dt} + i_L(t)R_1 - V + v_C(t) = 0 \qquad (4.16)$$

$$i_L(t) - C\frac{dv_C(t)}{dt} - \frac{v_C(t)}{R_2} = 0 \qquad (4.17)$$

7. We organize the n equations in the form $\dot{\mathbf{x}}(t) = \mathbf{A}\mathbf{x}(t) + \mathbf{B}\mathbf{u}(t)$

$$\begin{bmatrix} \dfrac{di_L(t)}{dt} \\ \dfrac{dv_C(t)}{dt} \end{bmatrix} = \begin{bmatrix} -\dfrac{R_1}{L} & -\dfrac{1}{L} \\ \dfrac{1}{C} & -\dfrac{1}{CR_2} \end{bmatrix} \begin{bmatrix} i_L(t) \\ v_C(t) \end{bmatrix} + \begin{bmatrix} \dfrac{1}{L} \\ 0 \end{bmatrix} [V] \tag{4.18}$$

$$\mathbf{A} = \begin{bmatrix} -\dfrac{R_1}{L} & -\dfrac{1}{L} \\ \dfrac{1}{C} & -\dfrac{1}{CR_2} \end{bmatrix} \tag{4.19}$$

$$\mathbf{B} = \begin{bmatrix} \dfrac{1}{L} \\ 0 \end{bmatrix} \tag{4.20}$$

8. We assume that the scope of our simulation is to calculate the current in the resistor R_2.
9. We write one algebraic equations that relates $i_{R2}(t)$ to the selected state variables $(i_L(t),\ v_C(t))$, the circuit input V and the circuit parameters (R_1, R_2, L, C):

$$i_{R2}(t) = \frac{v_C(t)}{R_2} \tag{4.21}$$

10. We organize the output equation in the form $\mathbf{y}(t) = \mathbf{C}\mathbf{x}(t) + \mathbf{D}\mathbf{u}(t)$:

$$[i_{R2}(t)] = \begin{bmatrix} 0 & \dfrac{1}{R_2} \end{bmatrix} \begin{bmatrix} i_L(t) \\ v_C(t) \end{bmatrix} + [0][V] \tag{4.22}$$

$$\mathbf{C} = \begin{bmatrix} 0 & \dfrac{1}{R_2} \end{bmatrix} \tag{4.23}$$

$$\mathbf{D} = [0] \tag{4.24}$$

Summarizing, the state-space model of the circuit of Figure 4.1 is given by the two equations:

$$\begin{bmatrix} \dfrac{di_L(t)}{dt} \\ \dfrac{dv_C(t)}{dt} \end{bmatrix} = \begin{bmatrix} -\dfrac{R_1}{L} & -\dfrac{1}{L} \\ \dfrac{1}{C} & -\dfrac{1}{CR_2} \end{bmatrix} \begin{bmatrix} i_L(t) \\ v_C(t) \end{bmatrix} + \begin{bmatrix} \dfrac{1}{L} \\ 0 \end{bmatrix} [V] \tag{4.25}$$

$$[i_{R2}(t)] = \begin{bmatrix} 0 & \dfrac{1}{R_2} \end{bmatrix} \begin{bmatrix} i_L(t) \\ v_C(t) \end{bmatrix} + [0][V] \tag{4.26}$$

Example 4.2 Let us now consider a more complicated case (Figure 4.2). Since the main challenges are related to the creation of (4.4), in this example we will focus on steps from 1 to 7.

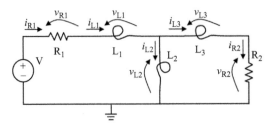

Figure 4.2   Linear dynamic circuit

1. In the circuit, there are three energy storage elements, three inductors, but the three inductors currents are linear dependent. This can be easily verified looking at the KCL:

$$i_{L1}(t) - i_{L2}(t) - i_{L3}(t) = 0 \qquad (4.27)$$

Thus, the system is of the second order, $n = 2$.

2. As state variables, we can select any two of the three inductor currents. For example, we decide to select $i_{L1}(t)$ and $i_{L3}(t)$.

3. We proceed writing two mesh equation using KVL, one for each inductor:

$$V_1 - v_{R1}(t) - v_{L1}(t) - v_{L2}(t) = 0 \qquad (4.28)$$

$$v_{L2}(t) - v_{L3}(t) - v_{R2}(t) = 0 \qquad (4.29)$$

4. We have two KVL that equal the number of state variables n.

5. Using the needed algebraic equations we re-arrange the n equations so that the only terms involved are the state variables and their derivatives $\left(i_{L1}(t),\ i_{L3}(t), \frac{di_{L1}(t)}{dt}, \frac{di_{L3}(t)}{dt} \right)$, the circuit inputs V and the circuit parameters $(R_1, R_2, L_1, L_3, L_2)$:

$$V_1 - R_1 i_{L1}(t) - L_1 \frac{di_{L1}(t)}{dt} - L_2 \frac{di_{L1}(t)}{dt} + L_2 \frac{di_{L3}(t)}{dt} = 0 \qquad (4.30)$$

$$L_2 \frac{di_{L1}(t)}{dt} - L_2 \frac{di_{L3}(t)}{dt} - L_3 \frac{di_{L3}(t)}{dt} - R_2 i_{L3}(t) = 0 \qquad (4.31)$$

6. We can now organize the n equations in the form $\dot{x}(t) = \mathbf{A}x(t) + \mathbf{B}u(t)$:

$$\begin{bmatrix} \dfrac{di_{L1}(t)}{dt} \\[2ex] \dfrac{di_{L3}(t)}{dt} \end{bmatrix} = A \begin{bmatrix} i_{L1}(t) \\[1ex] i_{L3}(t) \end{bmatrix} + B[V] \qquad (4.32)$$

$$A = \begin{bmatrix} \dfrac{L_2}{L_2 L_2 - L_1(L_3 + L_2) - L_2(L_3 + L_2)} R_1 & \dfrac{(L_3 + L_2)}{L_2} R_1 & \dfrac{L_2}{L_2 L_2 - L_1(L_3 + L_2) - L_2(L_3 + L_2)} R_2 & \dfrac{L_2 L_2}{L_2 L_2} R_2 \\ \dfrac{L_2 L_2}{L_2 L_2 - L_2(L_1 + L_2) - L_3(L_1 + L_2) L_2} R_1 & & \dfrac{L_2 L_2}{L_2 L_2 - L_2(L_1 + L_2) - L_3(L_1 + L_2) L_2} & \dfrac{(L_1 + L_2)}{L_2 L_2} R_2 \end{bmatrix}$$

$$B = \begin{bmatrix} -\dfrac{L_2}{L_2 L_2 - L_1(L_3 + L_2) - L_2(L_3 + L_2)} & \dfrac{(L_3 + L_2)}{L_2} \\ -\dfrac{L_2 L_2}{L_2 L_2 - L_2(L_1 + L_2) - L_3(L_1 + L_2) L_2} & \dfrac{1}{} \end{bmatrix}$$

When more than one derivative is included in each equation the organization of the *n* equations in the form $\dot{x}(t) = Ax(t) + Bu(t)$ can be long and error prone, in these cases it is more convenient to organize the *n* equations in the form of (4.33):

$$E\dot{x}(t) + Fx(t) + Gu(t) = 0 \qquad (4.33)$$

And after obtain matrix **A** and **B** as:

$$A = -E^{-1}F \qquad (4.34)$$

$$B = -E^{-1}G \qquad (4.35)$$

For this specific example (4.33) looks like:

$$\begin{bmatrix} -L_1 - L_2 & L_2 \\ L_2 & -L_2 - L_3 \end{bmatrix} \begin{bmatrix} \dfrac{di_{L1}(t)}{dt} \\ \dfrac{di_{L3}(t)}{dt} \end{bmatrix} + \begin{bmatrix} -R_1 & 0 \\ 0 & -R_2 \end{bmatrix} \begin{bmatrix} i_{L1}(t) \\ i_{L3}(t) \end{bmatrix} + \begin{bmatrix} 1 \\ 0 \end{bmatrix} [V] = 0$$

$$(4.36)$$

$$E = \begin{bmatrix} -L_1 - L_2 & L_2 \\ L_2 & -L_2 - L_3 \end{bmatrix} \qquad (4.37)$$

$$F = \begin{bmatrix} -R_1 & 0 \\ 0 & -R_2 \end{bmatrix} \qquad (4.38)$$

$$G = \begin{bmatrix} 1 \\ 0 \end{bmatrix} \qquad (4.39)$$

And so matrix **A** and **B** can be obtained as:

$$A = \begin{bmatrix} -L_1 - L_2 & L_2 \\ L_2 & -L_2 - L_3 \end{bmatrix}^{-1} \begin{bmatrix} -R_1 & 0 \\ 0 & -R_2 \end{bmatrix} \qquad (4.40)$$

$$B = \begin{bmatrix} -L_1 - L_2 & L_2 \\ L_2 & -L_2 - L_3 \end{bmatrix}^{-1} \begin{bmatrix} 1 \\ 0 \end{bmatrix} \qquad (4.41)$$

4.3 Discretization

Until now we have focused our attention on continuous time state-space model as in (4.13), but to execute the model on a computer we need to discretize it and obtain what is normally referred to as a discrete time state-space model:

$$x(t_c + h) = A_{disc}x(t_c)+B_{disc}u(t_c + h) \tag{4.42}$$

$$y(t_c + h) = C_{disc}x(t_c + h) + D_{disc}u(t_c + h) \tag{4.43}$$

Since (4.43) does not contain any derivative terms

$$C_{disc} = C \tag{4.44}$$

$$D_{disc} = D \tag{4.45}$$

Vice versa to obtain (4.42), (4.4) needs to be discretized to obtain A_{disc} and B_{disc}. It should be observed that the time instant to which the input vector refers is determined by the specific integration method applied.

Let us now proceed discretizing (4.4) using Euler forward, Euler backward, and trapezoidal rule so to calculate A_{EF}, B_{EF}, A_{EB}, B_{EB}, A_{TR}, B_{TR}.

Euler forward
Applying Euler forward integration method (4.46) to (4.4), we obtain (4.47):

$$x(t_c + h) = x(t_c)+h\dot{x}(t_c) \tag{4.46}$$

$$x(t_c + h) = x(t_c)+h(Ax(t) + Bu(t)) \tag{4.47}$$

Rearranging (4.47), we obtain (4.48) and so A_{EF}, (4.49), and B_{EF}, (4.50). I in (4.48) is an identity matrix with size equal to the order of the system:

$$x(t_c + h) = (I + hA)x(t_c)+hBu(t) \tag{4.48}$$

$$A_{EF} = (I + hA) \tag{4.49}$$

$$B_{EF} = hB \tag{4.50}$$

Euler backward
Applying Euler backward integration method (4.51) to (4.4), we obtain (4.52):

$$x(t_c + h) = x(t_c)+h\dot{x}(t_c + h) \tag{4.51}$$

$$x(t_c + h) = x(t_c)+h(Ax(t_c + h) + Bu(t_c + h)) \tag{4.52}$$

Rearranging (4.52), we obtain (4.53) and then (4.54):

$$(I - hA)x(t_c + h) = x(t_c)+hBu(t_c + h) \tag{4.53}$$

$$x(t_c + h) = (I - hA)^{-1}x(t_c)+(I - hA)^{-1}hBu(t_c + h) \tag{4.54}$$

Looking at (4.54), we obtain A_{EB}, (4.55), and B_{EB}, (4.56). It is important to note that in (4.54), the input vector refers to the time instant $t_c + h$:

$$A_{EB} = (I - hA)^{-1} \tag{4.55}$$

$$B_{EB} = (I - hA)^{-1}hB \tag{4.56}$$

Trapezoidal rule
Applying trapezoidal rule integration method (4.57) to (4.4), we obtain (4.58):

$$x(t_c + h) = x(t_c) + \frac{h}{2}[\dot{x}(t_c) + \dot{x}(t_c + h)] \tag{4.57}$$

$$x(t_c + h) = x(t_c) + \frac{h}{2}[Ax(t_c) + Bu(t_c) + Ax(t_c + h) + Bu(t_c + h)] \tag{4.58}$$

Rearranging (4.58), we obtain (4.59) and then (4.60):

$$\left(I - \frac{h}{2}A\right)x(t_c + h) = \left(I + \frac{h}{2}A\right)x(t_c) + \frac{h}{2}Bu(t_c) + \frac{h}{2}Bu(t_c + h) \tag{4.59}$$

$$x(t_c + h) = \left(I - \frac{h}{2}A\right)^{-1}\left(I + \frac{h}{2}A\right)x(t_c) + \left(I - \frac{h}{2}A\right)^{-1}\frac{h}{2}B[u(t_c) + u(t_c + h)] \tag{4.60}$$

Looking at (4.60), we obtain A_{TR}, (4.61), and B_{TR}, (4.62). It is important to note that in (4.60), the effective input vector that multiply the matrix B_{TR} is the sum of the input vector at the time instant t_c and at the time instant $t_c + h$:

$$A_{TR} = \left(I - \frac{h}{2}A\right)^{-1}\left(I + \frac{h}{2}A\right) \tag{4.61}$$

$$B_{TR} = \left(I - \frac{h}{2}A\right)^{-1}\frac{h}{2}B \tag{4.62}$$

Example 4.3 Let us consider the circuit of Example 4.1 and let us assume the parameter of Table 4.1.

Table 4.1 Parameters

R_1	1 Ω
R_2	10 Ω
L	1 mH
C	2 mF
V	10 V

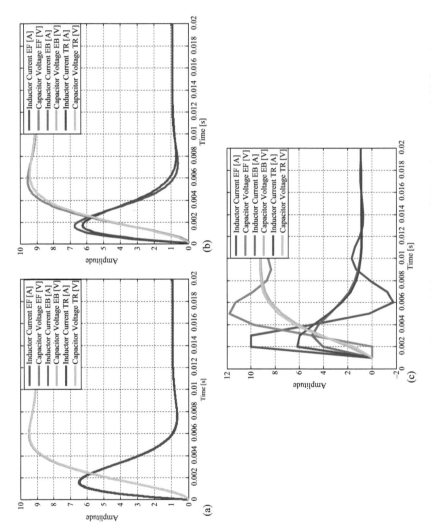

Figure 4.3   Example 4.3 simulation, (a) h = 0.01 ms, (b) h = 0.1 ms, and (c) h = 1 ms

In Figure 4.2, simulation results using the three integration method just discussed are reported. The results are consistence with the consideration reported in Chapter 1.

4.4 Automated state-space modeling

As mentioned before the automatic creation—based on topology and individual components models—of state-space system level models is not always possible. At the same time—assuming a certain structure for individual components—an approach was defined in [3] for automatic creation of state-space models. We proceed here describing this approach.

Before proceeding with the description of the methods itself, it is important to review a few concepts of circuit theory that are required to present the approach proposed in [1]. Let us start considering the example circuit of Figure 4.4. It is possible to write a matrix form of the Kirchhoff's current equations as in (4.63), where i_{br} is the vector of the currents flowing in each branch and A_a is the circuit incidence matrix. Let us start numbering the nodes and branches of the circuit of Figure 4.4, we can then initialize matrix A_a with a number of rows equal to the number of nodes and a number of columns equal to the number of branches. We then need to arbitrary orient the branches of the circuit; after that A_a can be populated, each column has exactly two nonzero elements, one equal to 1, the other to -1 depending on the defined branch orientation.

$$A_a i_{br} = 0 \qquad (4.63)$$

It is possible to linearly manipulate matrix A_a so to obtain a new matrix $\tilde{A}_a$ that has the characteristic shape of (3.65). It is so possible to write the matrix form of

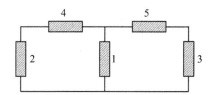

Figure 4.4   Generic circuit with five branches

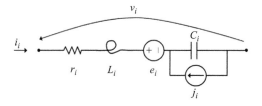

Figure 4.5   Prototypical branch

Kirchhoff's current law equations as in (4.66) where the matrix $\tilde{A}$ is obtained—as indicated in (4.65)—removing the zero row of matrix $\tilde{A}_a$. The fact that by manipulating A_a we obtain a new matrix that has a row of zeros should not surprise us: the matrix A_a has as many rows as node in the circuit but we know—as already discussed in Chapter 3—that only $n-1$ linear independent Kirchhoff's current law equations can be written for a given circuit:

$$\tilde{A}_a = \begin{bmatrix} I_{n-1,n-1} & \hat{A}_{n-1,b-n+1} \\ 0_{1,n-1} & 0_{1,b-n+1} \end{bmatrix} \tag{4.64}$$

$$\tilde{A} = \begin{bmatrix} I_{n-1,n-1} & \hat{A}_{n-1,b-n+1} \end{bmatrix} \tag{4.65}$$

$$\tilde{A} i_{br} = 0 \tag{4.66}$$

It is important to notice that the vector i_{br} of (4.66) is not the same vector i_{br} of (4.63) as during the process of creating $\tilde{A}_a$ rows re-ordering may have been necessary. For the sake of simplicity, we prefer to do not introduce a new vector and we will keep referring to i_{br}, assuming than this still represents the branches currents vector after manipulation of matrix A_a.

Looking at (4.66) is possible to separate the vector i_{br} in two components i_y and i_x where i_y can be seen as a vector of dependent current as they can be calculated from i_x using (4.68):

$$\begin{bmatrix} I & \hat{A} \end{bmatrix} \begin{bmatrix} i_y \\ i_x \end{bmatrix} = 0 \tag{4.67}$$

$$i_y = -\hat{A} i_x \tag{4.68}$$

The vector of branch currents i_{br} can then be expressed in function of the vector of independent currents i_x. The construction of the matrix B_b is the goal of this first introductory circuit review, it is worth to highlight that it can also be demonstrated that B_b satisfy a generalized form of Kirchhoff's voltage equation as in (4.70):

$$i_{br} = \begin{bmatrix} i_y \\ i_x \end{bmatrix} = \begin{bmatrix} -\hat{A} \\ I \end{bmatrix} i_x = B_b{}^T i_x \tag{4.69}$$

$$B_b v_{br} = 0 \tag{4.70}$$

We can now proceed analyzing in the detail the method presented in [1]. Let us define the prototypical circuit of figure for a hypothetical branch of the circuit we want to model.

With reference to Figure 4.3, the characteristic equation of the branch can be obtained writing the Kirchhoff's voltage equation across the branch and the characteristic equation of each individual component so obtaining the equation of (4.71):

$$v_i = r_i i_i + \frac{dL_i i_i}{dt} + P_i \int (i_i + j_i) + e_i \tag{4.71}$$

where P_i is the reciprocal of the branch capacitance C_i.

If we now assume that the circuit of interest is composed of branches that can be represented using the prototypical branch of Figure 4.3, we can express the vector of branch voltages v_{br} as in (4.72):

$$v_{br} = r_{br}i_{br} + \frac{dL_{br}i_{br}}{dt} + P_{br}q_{br} + e_{br} \tag{4.72}$$

where r_{br} is a diagonal matrix with individual branch resistance values on the diagonal, similarly L_{br} is a matrix with individual branch inductance values on the diagonal and eventually with mutually coupling inductance off the diagonal, P_{br} is a diagonal matrix with the reciprocal of the branch capacitance values on the diagonal, e_{br} is a vector populated with the values of individual branch voltage sources. q_{br} is obtained by (4.73), where j_{br} is a vector of the independent current sources connected at each branch:

$$q_{br} = \int (i_{br} + j_{br}) \tag{4.73}$$

Our goal now is to manipulate the system of equations of (4.72)—expressed in terms of branch voltages and currents—so to obtain a system of equations expressed in terms of a set of independent inductor currents and independent capacitor voltages.

We can first proceed focusing on the inductor currents, let us so left multiply both members of (4.72) for B_b and substitute i_{br} with $B_b^T i_x$. Remembering that $B_b v_{br} = 0$, we than obtain (4.74). As can be seen, the new obtained equation is expressed in terms of i_x instead of i_{br}:

$$0 = B_b r_{br} B_b^T i_x + B_b \frac{dL_{br}B_b^T i_x}{dt} + B_b P_{br} q_{br} + B_b e_{br} \tag{4.74}$$

Let us define a vector of capacitor currents i_C and a vector q_c as $q_c = \int i_C$ and a matrix M—its size is determined by the number of capacitor and by the number of branches—such that $m_{ij} = 1$ if the capacitor i is connected to the node j and zero otherwise. We can so write (4.75):

$$P_{br}q_{br} = P_{br}Mq_c \tag{4.75}$$

If we now substitute (4.75) in (4.74), we obtain (4.76), and further expanding also (4.77). Equation (4.77) can be written in a more compact form as in (4.78):

$$0 = B_b r_{br} B_b^T i_x + B_b \frac{dL_{br}B_b^T i_x}{dt} + B_b P_{br} Mq_c + B_b e_{br} \tag{4.76}$$

$$0 = B_b r_{br} B_b^T i_x + B_b \frac{dL_{br}}{dt} B_b^T i_x + B_b L_{br} B_b^T \frac{di_x}{dt} + B_b P_{br} Mq_c + B_b e_{br} \tag{4.77}$$

$$0 = r_x i_x + \frac{dL_x}{dt} i_x + L_x \frac{di_x}{dt} + P_x q_c + B_b e_{br} \tag{4.78}$$

where $r_x = B_b r_{br} B_b^T$, $L_x = B_b L_{br} B_b^T$, $P_x = B_b P_{br} M$.

Let us now focus on the capacitor voltages—more correctly charge—starting from (4.79)—that is obtained directly from (4.75). From (4.79), we can obtain (4.80):

$$q_c = M^T q_{br} \qquad (4.79)$$

$$\frac{dq_c}{dt} = M^T \frac{dq_{br}}{dt} \qquad (4.80)$$

Remembering of (4.73), we can rewrite (4.80) so to get rid of the term q_{br} and obtain (4.81)

$$\frac{dq_c}{dt} = M^T B_b{}^T i_x + M^T j_{br} \qquad (4.81)$$

We now have two systems of equations (4.78) and (4.81) expressed in terms of state variables q_c, i_x, of circuit parameters and of independent current and voltage sources j_{br}, e_{br}. We can rearrange them in the traditional form for LTI system so obtaining equations of (4.82). Matrices A and B are highlighted in (4.83) and (4.84):

$$\frac{d}{dt}\begin{bmatrix} q_c \\ i_x \end{bmatrix} = \begin{bmatrix} 0 & M^T B_b{}^T \\ -L_x{}^{-1} & -L_x{}^{-1}\left(r_x + \frac{dL_x}{dt}\right) \end{bmatrix}\begin{bmatrix} q_c \\ i_x \end{bmatrix} + \begin{bmatrix} M^T & 0 \\ 0 & -L_x{}^{-1} B_b \end{bmatrix}\begin{bmatrix} j_{br} \\ e_{br} \end{bmatrix}$$

$$(4.82)$$

$$A = \begin{bmatrix} 0 & M^T B_b{}^T \\ -L_x{}^{-1} & -L_x{}^{-1}\left(r_x + \frac{dL_x}{dt}\right) \end{bmatrix} \qquad (4.83)$$

$$B = \begin{bmatrix} M^T & 0 \\ 0 & -L_x{}^{-1} B_b \end{bmatrix} \qquad (4.84)$$

We can now focus on deriving a set of output equation that provide the whole set of branch voltages and currents.

Branch currents i_{br} can be obtained from the independent current vector i_x using (4.85). Notice that (4.85) was derived earlier in the chapter, it is reported here just for convenience:

$$i_{br} = \begin{bmatrix} i_y \\ i_x \end{bmatrix} = \begin{bmatrix} -\widehat{A} \\ I \end{bmatrix} i_x = B_b{}^T i_x \qquad (4.85)$$

For the branch voltages, we will use (4.72) and proceed with substituting (4.73) and once again (4.85). We so obtain the desired output (4.86), matrices C and D are highlighted (4.87) and (4.88):

$$\begin{bmatrix} i_{br} \\ v_{br} \end{bmatrix} = \begin{bmatrix} 0 & B_b{}^T \\ P_{br}M - L_{br}B_b{}^T L_x{}^{-1}P_x & (r_{br} + pL_{br})B_b{}^T - L_{br}B_b{}^T L_x{}^{-1}(r_x + pL_x) \end{bmatrix}\begin{bmatrix} q_c \\ i_x \end{bmatrix}$$
$$+ \begin{bmatrix} 0 & 0 \\ 0 & I - L_{br}B_b{}^T L_x{}^{-1}B_b \end{bmatrix}\begin{bmatrix} j_{br} \\ e_{br} \end{bmatrix}$$

$$(4.86)$$

$$C = \begin{bmatrix} 0 & B_b{}^T \\ P_{br}M - L_{br}B_b{}^T L_x{}^{-1} P_x & (r_{br} + pL_{br})B_b{}^T - L_{br}B_b{}^T L_x{}^{-1}(r_x + pL_x) \end{bmatrix}$$

(4.87)

$$D = \begin{bmatrix} 0 & 0 \\ 0 & I - L_{br}B_b{}^T L_x{}^{-1} B_b \end{bmatrix}$$

(4.88)

Example 4.4 Let us now focus on what are the required steps to use the described method to model a simple electrical circuit. We should first focus on the creation of matrices A and B according to (4.83) and (4.84). To do that we first need to define the following matrices and vectors: r_{br}, L_{br}, P_{br}, e_{br}, j_{br}, B_b, M.

Considering the circuit of Figure 4.6 with the parameters of Table 4.2 the indicated numbering for the branch and the nodes the matrices r_{br}, L_{br} and P_{br} are obtained by inspection—according to what indicated previous in the text—and are reported in (4.89), (4.90), and (4.91).

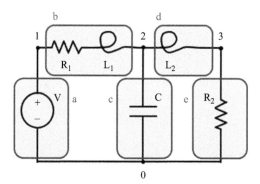

Figure 4.6   Example circuit for automated state-space modeling

Table 4.2   Circuit parameters

R_1	10 Ω
R_2	10 Ω
L_1	1 μH
L_2	10 mH
C	100 μF
V	10 V

$$r_{br} = \begin{bmatrix} 0 & 0 & 0 & 0 & 0 \\ 0 & R_1 & 0 & 0 & 0 \\ 0 & 0 & 0 & 0 & 0 \\ 0 & 0 & 0 & 0 & 0 \\ 0 & 0 & 0 & 0 & R_2 \end{bmatrix}$$

(4.89)

$$L_{br} = \begin{bmatrix} 0 & 0 & 0 & 0 & 0 \\ 0 & L_1 & 0 & 0 & 0 \\ 0 & 0 & 0 & 0 & 0 \\ 0 & 0 & 0 & L_2 & 0 \\ 0 & 0 & 0 & 0 & 0 \end{bmatrix} \tag{4.90}$$

$$P_{br} = \begin{bmatrix} 0 & 0 & 0 & 0 & 0 \\ 0 & 0 & 0 & 0 & 0 \\ 0 & 0 & 1/C & 0 & 0 \\ 0 & 0 & 0 & 0 & 0 \\ 0 & 0 & 0 & 0 & 0 \end{bmatrix} \tag{4.91}$$

Also by inspection, we obtain the vector e_{br}, j_{br} and M, reported in (4.92), (4.93), and (4.94):

$$e_{br} = \begin{bmatrix} V \\ 0 \\ 0 \\ 0 \\ 0 \end{bmatrix} \tag{4.92}$$

$$j_{br} = \begin{bmatrix} 0 \\ 0 \\ 0 \\ 0 \\ 0 \end{bmatrix} \tag{4.93}$$

$$M = \begin{bmatrix} 0 \\ 0 \\ 1 \\ 0 \\ 0 \end{bmatrix} \tag{4.94}$$

The matrix B_b can be calculated either by inspection or from the matrix A_a also obtained by inspection. Both matrices are reported in (4.95) and (4.96) for convenience:

$$B_b = \begin{bmatrix} 1 & -1 & -1 & 0 & 0 \\ 0 & 0 & -1 & 1 & 1 \end{bmatrix} \tag{4.95}$$

$$A_a = \begin{bmatrix} -1 & 0 & -1 & 0 & -1 \\ 1 & 1 & 0 & 0 & 0 \\ 0 & -1 & 1 & 1 & 0 \\ 0 & 0 & 0 & -1 & 1 \end{bmatrix} \tag{4.96}$$

L_x, r_x are obtained using equations $r_x = B_b r_{br} B_b{}^T$, $L_x = B_b L_{br} B_b{}^T$, as previously indicated in the text. We now have all the quantities needed to create the matrices A and B. To finalize the state-space model, we need to create also matrices

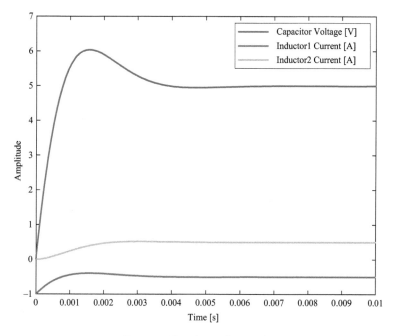

Figure 4.7 Example results

C and D. To calculate these two matrices, we only miss P_x, that is obtained using equation $P_x = B_b P_{br} M$. In Figure 4.7, we report results from the simulation of the derived model. Euler backward and time step of $1\,\mu s$ have been used for the numerical integration.

4.5 Simulation of state-space model

As mentioned in the introduction, state-space representation is a very common and useful modeling method that is widely used for simulation purpose. From the simulation point of view, the main advantages of this modeling method are:

- Often state-space modeling requires a significant less number of variable compared to nodal methods. Reducing the number of variable calculated at each time step reduce the simulation execution time and so the engineering time.
- If the matrices A and B are discretized using an explicit method the computational cost to solve (4.46) grows with the second power of the system size, $O(n^2)$, while in nodal method, the computational cost to solve (2.5) grows with the cube of the system size, $O(n^3)$, independently of the integration method used.
- If the matrices A and B are discretized using an explicit method, the solution of (4.46) can be fully parallelized.

- The modeling is done in continuous time while discretization is applied once the whole model is created. This can be advantageous since different discretization approaches can be easily compared.

At the same time, it is important to notice that in the case of linear time invariant systems, the computational cost to solve (3.5) grows with the quadratic of the system size, $O(n^2)$, since conductance matrix factorization is performed off-line. Moreover, electrical system typically requires implicit integration and, if this is the case, a good part of the advantages of state-space-based simulation are lost.

In any case, the most common way to simulate a system modeled using a state-space representation is not to directly integrate a system representation in the form of (4.13) but to use what is normally referred as a signal solver. In a signal solver, the system modeled using state-space representation is represented by a block diagram, the solution is computed by solving each block independently. For each block, the output is computed starting from its inputs values obtained from the solution at the previous time step. It is clear that this solution process implies the use of an explicit integration method. It is worth to mention that implicit or semi-implicit integration methods have been applied to signal flow solver but their use introduces additional challenges and it is not approached in those notes.

4.6   Signal flow solver

The signal flow model of a system can be obtained in different ways. Let us start assuming we decide to create such a model from a state-space representation in the form of (4.13). In this case, the signal flow model has a peculiar structure: we will analyze here the case of a system with one input and two state variables but the flow diagrams of larger systems present significant similarity.

Let us so consider the second-order system with one input of (4.97)

$$\dot{x}(t) = \begin{bmatrix} a_{11} & a_{12} \\ a_{21} & a_{22} \end{bmatrix} x(t) + \begin{bmatrix} b_1 \\ b_2 \end{bmatrix} u(t) \tag{4.97}$$

If we proceed integrating both the side of the system of equations and separating the two equations, we obtain:

$$x_1(t) = \int [a_{11}x_1(t) + a_{12}x_2(t) + b_1 u(t)] \tag{4.98}$$

$$x_2(t) = \int [a_{21}x_1(t) + a_{22}x_2(t) + b_2 u(t)] \tag{4.99}$$

The signal flow diagram of the system described by (4.98) and (4.99), Figure 4.4, can be created using three basic blocks: "sum," "gain," and "integrator."

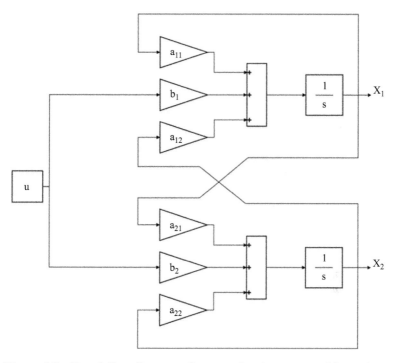

Figure 4.8 Signal flow diagram of a second-order system with one input

Example 4.5 Let us assume that we want to create the signal flow diagram of the system modeled in Examples 4.1 and 4.2. Since both the systems are of the second order with only one input, the flow diagram has the structure of the one of Figure 4.8. The gain coefficients clearly change according with the specific example.

In the case of Example 4.1, we have:

$$a_{11} = -\frac{R_1}{L}, \quad a_{12} = -\frac{1}{L}, a_{21} = \frac{1}{C}, a_{11} = -\frac{1}{CR_2}, b_1 = \frac{1}{L}, b_2 = 0$$

In the case of Example 4.2, we have:

$$a_{11} = \frac{L_2}{L_2 L_2 - L_1(L_3 + L_2) - L_2(L_3 + L_2)} \frac{(L_3 + L_2)}{L_2} R_1,$$

$$a_{12} = \frac{L_2}{L_2 L_2 - L_1(L_3 + L_2) - L_2(L_3 + L_2)} R_2$$

$$a_{21} = \frac{L_2 L_2}{L_2 L_2 - L_2(L_1 + L_2) - L_3(L_1 + L_2)} \frac{R_1}{L_2},$$

$$a_{11} = \frac{L_2 L_2}{L_2 L_2 - L_2(L_1 + L_2) - L_3(L_1 + L_2)} \frac{(L_1 + L_2)}{L_2 L_2} R_2$$

$$b_1 = -\frac{L_2}{L_2L_2 - L_1(L_3 + L_2) - L_2(L_3 + L_2)} \cdot \frac{(L_3 + L_2)}{L_2},$$

$$b_2 = -\frac{L_2L_2}{L_2L_2 - L_2(L_1 + L_2) - L_3(L_1 + L_2)} \frac{1}{L_2}$$

Another and often more convenient way to create a signal flow representation of a system is to transform in signal flow the state equations describing the system of interest before obtaining the representation of (4.13).

Example 4.6 Let us consider Example 4.2. A signal flow diagram can be obtained directly working on (4.31) and (4.32). Integrating both the equations and rearranging the terms, we obtain:

$$i_{L1}(t) = \frac{1}{L_1 + L_2} \int [-R_1 i_{L1}(t) + V_1] + \frac{L_2}{L_1 + L_2} i_{L3}(t) \tag{4.100}$$

$$i_{L3}(t) = \frac{1}{L_2 + L_3} \int [-R_2 i_{L3}(t)] + \frac{L_2}{L_2 + L_3} i_{L1}(t) \tag{4.101}$$

Equations (4.100) and (4.101) lead to the flow diagram of Figure 4.9.

One issue of the approach we followed until now is that for each system we want to analyze we have to create a new state-space representation starting from scratch, despite the fact that different systems may be composed of the same basic

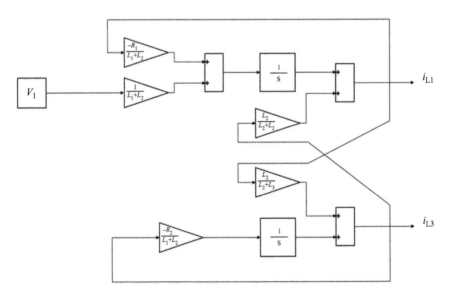

Figure 4.9   Flow diagram of Example 4.2

elements e.g. voltage sources, inductors, capacitors. To overcome this limitation, another possible approach to create signal flow models is to interconnect simple I/O models, each representing a single component, using KVL and KCL. This approach is clearly advantageous because it allows capitalizing past modeling effort without starting from scratch each time, at the same time, present some significant problems that are the main drawbacks of signal flow solvers if compared to nodal ones.

The main problem is that we want to define inputs and outputs for systems that in reality do not have inputs and outputs. Let us for example consider a resistor, either assuming the voltage as an input and the current as an output either assuming the current as an input and the voltage as an output is somehow wrong: a resistor define a relation between voltage and current at its terminals. This modeling "error" has two main effects. First, while obtaining the system equation by manual interconnection of basic components (manually imposing KVL and KCL), it is easy to create simulation artifacts that lead to violations of the conservation of energy principle. Second, it is often necessary to have multiple models of the same component e.g. a resistor model that assume current as an input and voltage as an output and one that assume voltage as an input and current as an output; while this can be seen as an easily surmountable problem that only requires a larger components library, it is worth to mention that in some cases, a specific selection of input and output may be particularly problematic. Let us consider for example the case of an ideal inductor, selecting the current as an input and the voltage as an output implies the use of a pure differentiator that may lead to numerical problem, vice versa assuming the voltage as an input and the current as an output imply the use of an integrator that it is always a much safer choice.

It is worth to underline how vice versa the interconnection of components in nodal analysis is based on a principle called natural coupling. One component model in resistive companion has multiple terminals and two variables, voltage and current, are associated with each terminal. The connection of two terminals implies the conservation of those quantities and so the conservation of their product: power.

Let us assume we create a basic library of components consisting of: voltage source, resistor, and inductor.

For the voltage source, we have a model consisting of a constant with an output equal to the voltage defined for the specific voltage source (Figure 4.10).

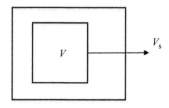

Figure 4.10   Voltage source signal flow model

For the resistor, we assume as an input the current and as an output the voltage, as a consequence, the resistor model consists of a gain element with value equal to the resistance defined for the specific resistor (Figure 4.11).

For the inductor, we assume as an input the voltage and as an output the current, as a consequence, the resistor model consists of an integrator and of a gain element with value equal to the inverse of the inductance defined for the specific inductor (Figure 4.12).

Let us now assume we want to create the signal flow model of the circuit of Figure 4.13.

The signal flow model, Figure 4.14, of this circuit can be created using the components model we just create satisfying the KVL of (4.102) and the KCL of (4.103):

$$V - v_{R1} = v_{L1} \tag{4.102}$$

$$i_{R1} = i_{L1} \tag{4.103}$$

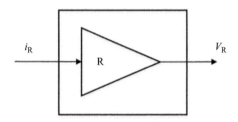

Figure 4.11   Resistor signal flow model

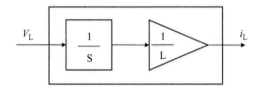

Figure 4.12   Inductor signal flow model

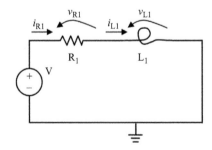

Figure 4.13   Example 4.6 Circuit 1

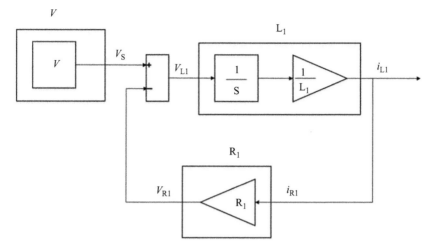

Figure 4.14 Signal flow diagram of the RL circuit of Figure 4.8

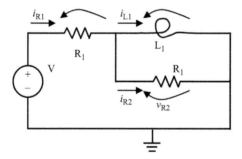

Figure 4.15 Example 4.6 Circuit 2

Let us now assume we want to create the signal flow model of the slight modified circuit of Figure 4.15.

In this case to create the signal flow model, Figure 4.16, of the considered circuit such that satisfies the KVL of (4.104) and the KCL of (4.105), we need to introduce a new model for resistor R_2. This model assumes the voltage as an input and the current as an output and is composed of a gain element with value equal to the inverse of the resistance defined for the specific resistor:

$$V - v_{R1} = v_{L1} = v_{R1} \tag{4.104}$$

$$i_{R1} = i_{L1} + i_{R2} \tag{4.105}$$

The new signal flow diagram is reported in Figure 4.16.

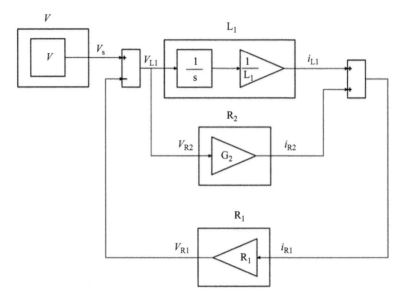

Figure 4.16   Signal flow diagram of the RL circuit of Figure 4.15

4.7   From state-space to transfer function representation

In many practical cases, for example for the design of a controller, it is convenient to transform a state-space model in a transfer function or in a set of transfer functions.

Let us start considering a first-order system with one input and one output. The state-space model looks like:

$$\dot{x}(t) = ax(t) + bu(t) \tag{4.106}$$

$$y(t) = cx(t) + du(t) \tag{4.107}$$

Applying Laplace transformation to (4.106) and (4.107), we obtain:

$$sX = aX + bU \tag{4.108}$$

$$Y = cX + dU \tag{4.109}$$

Rearranging (4.108), we obtain:

$$(s - a)X = bU \tag{4.110}$$

$$X = \frac{b}{(s - a)}U \tag{4.111}$$

Inserting (4.111) in (4.109), we obtain:

$$Y = \left(\frac{bc}{(s - a)} + d \right) U \tag{4.112}$$

And so the transfer function:

$$\frac{Y}{U} = \frac{sd + (bc - ad)}{(s - a)} \tag{4.113}$$

Example 4.7 Let us consider the example of (4.114) and (4.115)

$$\dot{x}(t) = -3x(t) + 2u(t) \tag{4.114}$$

$$y(t) = 2x(t) \tag{4.115}$$

The associated transfer function is:

$$\frac{Y}{U} = \frac{4}{s + 3} \tag{4.116}$$

Let us now consider the case of a generic system with n states m inputs and p outputs:

$$\dot{x}(t) = Ax(t) + Bu(t) \tag{4.117}$$

$$y(t) = Cx(t) + Du(t) \tag{4.118}$$

Following the same approach, we used for the single-input single-out first-order system, we can now precede transforming a generic n-order system with m inputs and p outputs:

$$sX = AX + BU \tag{4.119}$$

$$(sI - A)X = BU \tag{4.120}$$

$$X = (sI - A)^{-1}BU \tag{4.121}$$

$$Y = C(sI - A)^{-1}BU + DU \tag{4.122}$$

$$\frac{Y}{U} = C(sI - A)^{-1}B + D = G(s) \tag{4.123}$$

Example 4.8 Let us consider the second-order example of (4.124) and (4.125) with one input and one output:

$$\dot{x}(t) = \begin{bmatrix} 0 & 1 \\ -3 & -4 \end{bmatrix} x(t) + \begin{bmatrix} 0 \\ 1 \end{bmatrix} u(t) \tag{4.124}$$

$$y(t) = [5 \quad 1]x(t) + 0u(t) \tag{4.125}$$

Applying the formula of (4.123), we get:

$$G(s) = [5 \quad 1]\left(sI - \begin{bmatrix} 0 & 1 \\ -3 & -4 \end{bmatrix}\right)^{-1}\begin{bmatrix} 0 \\ 1 \end{bmatrix} \tag{4.126}$$

where

$$\left(sI - \begin{bmatrix} 0 & 1 \\ -3 & -4 \end{bmatrix}\right) = \begin{bmatrix} s & -1 \\ 3 & s+4 \end{bmatrix} \tag{4.127}$$

We have to calculate the inverse of this matrix analytically; since the matrix is a 2×2 we can do manually remembering the formula:

$$\begin{bmatrix} a & b \\ c & d \end{bmatrix}^{-1} = \frac{1}{ad-bc} \begin{bmatrix} d & -b \\ -c & a \end{bmatrix} \tag{4.128}$$

and so we have:

$$\begin{bmatrix} s & -1 \\ 3 & s+4 \end{bmatrix}^{-1} = \frac{1}{s(s+4)+3} \begin{bmatrix} s+4 & 1 \\ -3 & s \end{bmatrix} \tag{4.129}$$

And so plugging (4.129) into (4.126), we obtain:

$$\frac{1}{s(s+4)+3} \left(\begin{bmatrix} 5 & 1 \end{bmatrix} \begin{bmatrix} s+4 & 1 \\ -3 & s \end{bmatrix} \begin{bmatrix} 0 \\ 1 \end{bmatrix} \right) \tag{4.130}$$

The numerator of our transfer function given by equation

$$\begin{bmatrix} 5 & 1 \end{bmatrix} \begin{bmatrix} s+4 & 1 \\ -3 & s \end{bmatrix} \begin{bmatrix} 0 \\ 1 \end{bmatrix} = \begin{bmatrix} 5 & 1 \end{bmatrix} \begin{bmatrix} 1 \\ s \end{bmatrix} = 5+s \tag{4.131}$$

And so finally the transfer function can be obtained:

$$G(s) = \frac{s+5}{s^2+4s+3} \tag{4.132}$$

In the previous example, we still consider the particular case of a single-input single-output (SISO) systems, but the process just described can be used also for larger multi input multi output (MIMO) systems. For linear time invariant MIMO system, we can use a representation that extends the idea of a transfer function: the transfer matrix function:

$$\begin{bmatrix} Y_1 \\ Y_2 \\ \vdots \\ Y_p \end{bmatrix} = \begin{bmatrix} G_{1,1} & \cdots & G_{1,m} \\ \vdots & \ddots & \\ G_{p,1} & & G_{p,m} \end{bmatrix} \begin{bmatrix} U_1 \\ \vdots \\ U_m \end{bmatrix} \tag{4.133}$$

where $G_{p,m}$ expresses the relation between input m and output p.

Exercises

Exercise 1 Let us consider the circuit of Figure 4.17 and the parameter in Table 4.3.

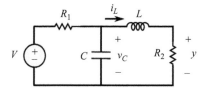

Figure 4.17 Circuit for exercise 1

Table 4.3 Circuit parameters

V	1 V
R_1, R_2	1 Ω
C	1 mF
L	1 mH
Δt	1 ms
$V_C(0)$	0 V
$I_L(0)$	0 A

(A) How many state variables are associated with the circuit?

(B) Could vector $x = \begin{bmatrix} u_C \\ u_{R2} \end{bmatrix}$ be an appropriate state vector?

(C) Derive the state-space model of the circuit.

(D) Use the trapezoidal rule to calculate one time step, using the given initial conditions.

Exercise 2 Let us consider the circuit of Figure 4.18 and the parameter in Table 4.4.

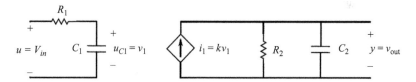

Figure 4.18 Circuit for exercise 2

Table 4.4 Circuit parameters

V_{in}	1 V
R_1, R_2	1 Ω
C_1, C_2	1 mF
k	10 A/V
Δt	1 ms
$V_{C_1}(0)$	0 V
$V_{C_2}(0)$	0 V

(A) Select a suitable state vector. Justify your answer.

(B) Could vector $x = \begin{bmatrix} i_1 \\ i_{R2} \end{bmatrix}$ be an appropriate state vector?

(C) Derive the state-space model of the circuit with . $u = V_{in}$ and $y = v_{out}$

(D) Use trapezoidal rule to calculate one time step, taking zero initial conditions and time $\Delta t = 1$ ms.

Exercise 3 Let us consider the circuit of Figure 4.19 and the parameter in Table 4.5.

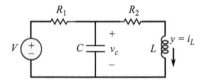

Figure 4.19   Circuit for exercise 2

Table 4.5   Circuit parameters

V	1 V
R_1, R_2	1 Ω
C	1 mF
L	1 mH
Δt	1 ms
$V_C(0)$	0 V
$I_L(0)$	0 V

(A) Select suitable state vector and derive state-space model for input $u = V$ and output $y = I_L$

(B) Use predictor–corrector method to calculate one simulation step.

(C) Use trapezoidal rule to calculate one simulation step.

(D) Solve the circuit by applying the modified nodal analysis method with a trapezoidal rule. Calculate one simulations step.

Exercise 4 Let us consider the circuit of Figure 4.20 and the parameter in Table 4.6.

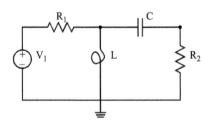

Figure 4.20   Circuit for exercise 2

Table 4.6   Circuit parameters

V	10 V
R_1, R_2	1 Ω
C	1 F
L	1 H
$V_C(0)$	0 V
$I_L(0)$	0 V

1. Which is the order of the system? How many input it has? Comment your answer.
2. Obtain (manually) a state-space representation of the circuit in the form:

$$\dot{x}(t) = Ax(t) + Bu(t)$$

3. Assuming that you variable of interest is the current is resistor R_1, write an output equation that give you I_{R1} and arrange it in the form:

$$y(t) = Cx(t) + Du(t)$$

4. Write a script (with a language of your choice) that simulates the circuit of Figure 4.1 on the base of the obtained description and plots I_{R1}. Compare the results obtained integrating the system of equation using Euler forward, Euler backward, trapezoidal rule for three different time step sizes: 0.1 s, 1 s, and 2 s. Select an appropriate final time for the simulation.
5. Write a script (with a language of your choice) that compares the results obtained modeling the circuit using resistive companion and using state-space representation. Use Euler backward as integration method and I_{R1} for the comparison. Comment the results obtained. Use a time step sizes equal to 0.1 s. Select an appropriate final time for the simulation.

Exercise 5 Let us consider the circuit of Figure 4.21 and the parameter in Table 4.7.

1. Derive the state-space model of the circuit using the method described in the paper: O. Wasynczuk, S.D. Sudhoff, "Automated state model generation algorithm for power circuits and systems", IEEE Transaction on Power Systems, Vol. 11, No. 4, 1996.

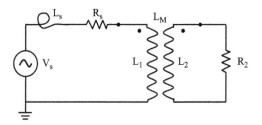

Figure 4.21   Circuit for exercise 2

Table 4.7   Circuit parameters

$|V_s|$	10 V
f	50 Hz
R_s	0.1 Ω
L_s	1 μH
L_1	1 mH
L_2	4 mH
L_M	2 mH
R_2	10 Ω

2. Plot the voltage and the current for each branch of the system. Select a proper time step and final time for the simulation. Use Euler backward for integration.

References

[1] F.E. Cellier and E. Kofman, *"Continuous System Simulation"*, New York, NY: Springer Science & Business Media, 2006.
[2] R.L. Williams II and D.A. Lawrence, *"Linear State-Space Control Systems"*, New York, NY: Wiley, 2007.
[3] O. Wasynczuk and S.D. Sudhoff, "Automated state model generation algorithm for power circuits and systems", *IEEE Transaction on Power Systems*, Vol. 11, No. 4, 1996.

Chapter 5

Parallelization methods

Andrea Benigni[1,2]

5.1 Introduction

The complexity of modern energy systems poses significant challenges on how these systems are planned, designed and operated. The design of each part (subsystem, component, algorithm) is a challenge due to the interactions and dimensions of the problem. This challenge cannot be simplified via de-coupling without risking the loss of essential dynamic behaviors. The impact of individual elements on the system, and vice versa, the impact of the system on individual elements, may not be inferred analytically, and the traditional design spiral maybe inadequate. In this context the use of numerical simulation tools becomes a critical need. Simulation is already a fundamental tool that supports design and analysis in many engineering fields. At the same time, despite a long history of use, traditional approaches and commercial tools are severely limited when treating complex, temporally and spatially distributed systems such as modern energy systems, partially because still largely based on serial execution algorithms. With time scales that span over 10 orders of magnitude and with systems of very large size that cannot be easily partitioned, high parallelizable simulation methods that ensure a computationally effective and scalable solution are needed.

In this chapter, we will review some of the most significant and recent methodologies that have been developed for the parallelization of dynamic time-domain simulation. Since the focus of this book is on electromagnetic transient simulation, from a mathematical point of view this means we will focus on the parallelization of the solution of a set of ordinary differential equations (ODE). Before moving forward it worth to spend a few words on the parallelization of transient stability simulation – that is the solution of a system of differential algebraic equations (DAE). The parallelization of transient stability simulation solvers has been widely investigated in the last 20 years: in [1–3], several theoretical approaches have been proposed, in [4,5], the use of a GPUs and of a cluster system have been investigated. In recent years, due to the increasing interest—among others also of the US Department of Energy—in dynamic security assessment, faster than real-time transient stability simulation [6] based on

[1]Institute of Energy and Climate Research: Energy Systems Engineering (IEK-10), Juelich Research Center, Germany
[2]Department of Mechanical Engineering, RWTH Aachen University, Germany

parallelization methods and on the use of high performance computing techniques [7,8] has received more and more attention. The main difference between the parallelization of transient stability simulation and the parallelization of electromagnetic transient simulation is in the quasi static assumption for the interconnections that is done in transient stability simulation. This assumption creates a first chance of parallelization decoupling the solution of the dynamic differential equations. Another important aspect to be considered is that transient stability simulation typically targets a time step size varying from tens of millisecond to hundreds of millisecond, while electro-magnetic transient simulation targets a time step size varying from tens of microsecond to tens of nanosecond; this creates different requirements and possibilities for the selection of the parallelization approach.

The methods reviewed in this chapter can be classified in two broad categories. The first one is the one of methods that decomposes the solution of the modeled system so that part of it can be parallel executed, without taking advantage of any dynamic characteristics of the system of interest. Those methods are typically characterized by a serial execution step, that is needed to recombine the system solution, but have the major advantage to provide an exact solution identical to the one provided by a tra-ditional serial execution based solver. For what concern this family of methods, we will discuss in detail—for it is historical importance—the method so called Diakoptics and a more recently introduced method that combine state space and nodal representation. The second category is the one of methods that parallelize simulation execution relying on the dynamic characteristics of the system of interest and exploiting different inte-gration schemes so to take advantage of the latency that characterize any dynamic system. Latency can be seen as system property where some quantities remain constant or change very slowly in comparison to the rest of system. From a discrete time simulation point of view, this means that a part of the system can be considered dor-mant and the related quantities constant for one or more time steps.

Before proceeding with a detailed description of the parallelization methods considered in this book chapter, we provide two case studies that anticipate how two of the methods presented in paper can be used for engineering activities.

5.2   Case study 1: parallelize the simulation of a ship power system

In this case study, we will show how the latency-based linear multi-step compound (LB-LMC) method—one of the methods described later in this chapter—has been used for real-time simulation of a naval ship power system and for the verification of power electronics controller operation. The LB-LMC method has been developed under support of the Office of Naval Research (ONR) to enable high-speed, scalable, real-time simulation of power electronics-based power systems, in particular the goal was to support the real-time simulation and Hardware In the Loop (HIL) testing of system based on silicon carbide high-power converters characterized by switching frequency in the 100–200 kHz range with time step smaller than 100 ns.

Results about the development and use of this method have been reported in several papers, for this case study we will refer to the system used in [9]. The testing scenario is

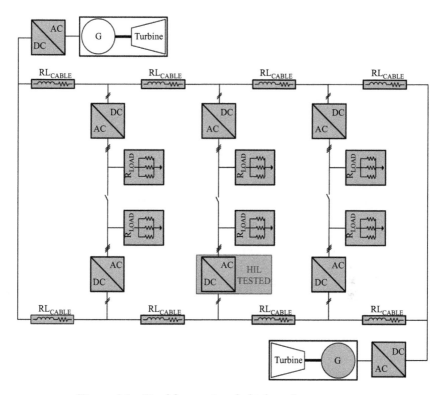

Figure 5.1 Dual-bus notional shipboard power system

based on the dual-bus notional shipboard power system displayed in Figure 5.1. The system is composed of six DC/AC converters and two DC/DC converters. Parameters for this system are set so to obtain 40 MW of total load and the DC/DC converters are set to output 12 kV DC voltage onto bus lines. The overall system has 54 nodes, and 96 switches. According with LB-LMC method, each non-linear component is modeled using a state-space approach and has been integrated using an explicit algorithm. The rest of the system, composed only of linear components, is modeled using resistive companion as introduced in Chapter 3. Using this approach, each non-linear component is solved in parallel and the solution of the linear part of the network can be highly optimized. Later in this chapter, we will provide more details on this method.

This method has been used in [10] the evaluate in HIL the control of one of the zonal converter of Figure 5.1. Five of the DC/AC converters are controlled internally so that the converters produce 3-phase AC output with a frequency of 60 Hz. The converter controllers are set to operate with 100 kHz switching frequency. Similarly, the DC/DC converters operate at a 100 kHz switching frequency. One of the DC/AC converters in Figure 5.1 is configured to operate with a closed-loop controller which is implemented externally from the model and FPGA that simulates said model. The three phases AC voltages of this converter are converted to analog signals via a digital/analog converter, so to be sampled by the external controller. Moreover, this converter is configured to take three switch control signals provided by the external controller. In this way, we verified the operation of a simple controller for a three-phase inverter.

While the HIL simulation was executing, modeling the shipboard system to run with 100 kHz switching frequency and 50 ns time step, the analog signals of the externally controlled converter's three phases were captured via an oscilloscope. During the capturing, the controller was set to change from maximum output amplitude (6 kV) to half output amplitude (3 kV). The real-time capture of the converter output during the control change is shown in Figure 5.2. Along with control change results, the 100 kHz switching ripple found in the converter output was also captured in real-time during another execution of the HIL simulation while control reference was held constant; captured result is shown in Figure 5.3.

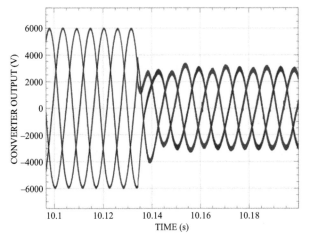

Figure 5.2   Converter output after control change

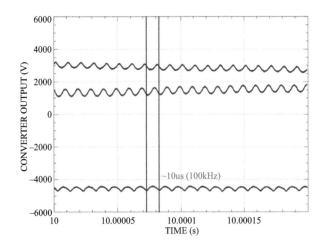

Figure 5.3   Converter output ripple

5.3   Case study 2: parallelize the simulation of the IEEE 34 and IEEE 123 distribution network

In this case study, we show how the state-space nodal method—first introduced in [11] and detailed described later in this chapter—has been used to parallelize a modified version of the IEEE34 and IEEE123 distribution network.

In [12], the author proposed an approach for decentralized load estimation for distribution systems using artificial neural networks and the proposed method have been tested using a HIL approach, a high level diagram of the experimental set-up is reported in Figure 5.4. Six low cost control platform based on a quad-core ARM processor are connected to a real-time simulator from the company Opal-RT through the use of analog signals so to receive power systems measurements, the control boards communicate between them using UDP. To evaluate the developed algorithms in a realistic environment, the communication network performances are emulated using a commercial network emulator from the company Apposite.

Both the IEEE34 and IEEE123 test feeder have been modified to include distributed generation and daily load variability. In the original IEEE34 node, the system is an actual feeder located in Arizona. It is an unbalanced system with both spot loads and distributed loads. In [12], the unbalanced loads in each three-phase sections are summed up and taken as total three-phase balanced loads and six PV generation units have been added. The IEEE123 node test feeder is configured so that the normally opened switch between the node 59 and node 121 is closed to create a weakly meshed distribution system and 6 PV generation units are added at

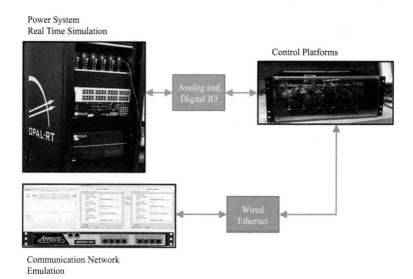

Figure 5.4   HIL platform

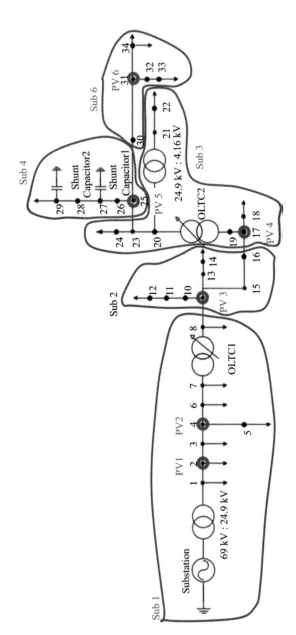

Figure 5.5 Partitioning of the IEEE34 distribution network

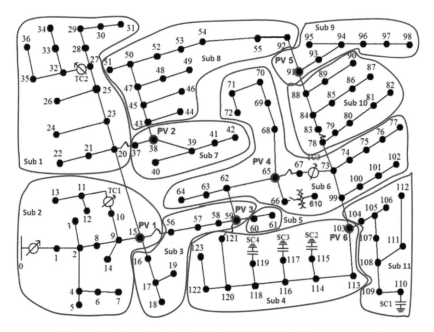

Figure 5.6 Partitioning of the IEEE123 distribution network

bus 15, 38, 59, 65, 91, and 103, with the nominal capacity of 400, 1200, 550, 350, 600, and 400 kVA, respectively.

Both the IEEE34 and the IEEE123 are too large to be executed with a 50-μs time step on a single core so it is execution need to be distributed on multiple cores. Using five SSN nodes, the original IEEE34 distribution network is divided in five subsystems (Figure 5.5), for the original IEEE123 seven SSN nodes have been used so to obtain 11 subsystems (Figure 5.6). Both IEEE test cases have been simulated using five cores and a 50-μs time step.

5.4 Diakoptics

The diakoptics approach has its origin in the method developed by Gabriel Kron in [13] and consist in dividing the network of interest in a set of sub-system, each sub-system conductance matrix is then inverted independently while a series of constrain are added to ensure that the solution is equivalent to the one of the original systems. The benefit of diakoptics arises from the fact that the solution of a linear system is proportional to the cube of the size of the system. As a consequence, if for example we partition a system in two sub-systems of the same size the computational cost reduce—without considering the computational cost deriving from the added con-strains—of about eight times. The key aspect of this approach is that in general in a network the elements are not completely interdependent, in the sense that one element is not directly linked to all the other elements in the network. As a consequence, a network can be divided in subsystems by removing a limited number of branches, this is important because as we will see later the number of constrain to be added—so the additional computation needed—depends on the number of removed branches.

The greatest advantage of diakoptics—if compared to other parallelization approaches that we will analyze later in this chapter—is that the solution so obtained is the exact solution of the original system without any additional approximation. Another advantage of diakoptics is that it does not require any specific dynamic or modeling characteristics for the system simulated.

We will proceed now describing how diakoptics is applied to nodal analysis using the same notation introduced in [14]. As described in [14], diakoptics can also be applied to mesh analysis. For the sake of simplicity, we will proceed describing the diakoptics approach applying it to a static network. It is worth to remind that using the resistive companion approach—as described in Chapter 3—diakoptics can be used to simulate dynamic networks.

Let us start considering the circuit of Figure 5.7. The circuit is composed of 5 nodes and 10 branches, we also assume that a current source is connected to each node. For reference, the solution of the system in Figure 5.7 can be obtained using nodal analysis, as for (5.1):

$$Y_{aa}v_a = I_a \tag{5.1}$$

where:

$$v_a = \begin{bmatrix} v_a & v_b & v_c & v_d & v_e \end{bmatrix}^T \tag{5.2}$$

$$I_a = \begin{bmatrix} I_a & I_b & I_c & I_d & I_e \end{bmatrix}^T \tag{5.3}$$

$$Y_{aa} = \begin{bmatrix} Y_1+Y_3+Y_j & -Y_1 & 0 & 0 & -Y_j \\ -Y_1 & Y_1+Y_4+Y_2 & -Y_2 & 0 & 0 \\ 0 & -Y_2 & +Y_5+Y_2+Y_k & -Y_k & 0 \\ 0 & 0 & -Y_k & +Y_7+Y_k+Y_6 & -Y_6 \\ -Y_j & 0 & 0 & -Y_6 & +Y_8+Y_6+Y_j \end{bmatrix} \tag{5.4}$$

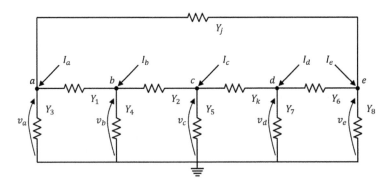

Figure 5.7   Original example circuit

Observing the structure of the conductance matrix can be seen that the system of Figure 5.1 can be separated in two subsystems by simply removing two branches, branch J and anyone between branch 1, 2, K and 6. Since it is convenient to separate the system in two subsystems of—as much as possible—the same size, the best choice is to remove branch 2 or branch k. Let us so precede removing branch J and K, we then obtain two separate subsystems, as in Figure 5.8. In place of the removed branches, we add a current source to each of the nodes where the branches were connected.

We can then precede applying nodal analysis to the two independent circuits obtaining (5.5) and (5.6):

$$Y_{11}v_1 = I_1 + \tilde{I}_1 \tag{5.5}$$

$$Y_{22}v_2 = I_2 + \tilde{I}_2 \tag{5.6}$$

where:

$$\tilde{I}_1 = \begin{bmatrix} \tilde{I}_a & 0 & \tilde{I}_c \end{bmatrix} \tag{5.7}$$

$$\tilde{I}_2 = \begin{bmatrix} \tilde{I}_d & \tilde{I}_e \end{bmatrix} \tag{5.8}$$

Equations (5.5) and (5.6) can be combined to obtain a system of (5.9) that describe the two independent subsystems:

$$\tilde{Y}_{aa}v_a = I_a + \tilde{I}_a \tag{5.9}$$

where:

$$\tilde{I}_a = \begin{bmatrix} \tilde{I}_a & 0 & \tilde{I}_c & \tilde{I}_d & \tilde{I}_e \end{bmatrix} \tag{5.10}$$

We can proceed looking at the branch that we removed. Each removed branch is connected to a voltage source, obtaining the circuit of Figures 5.9 and 5.10.

Applying mesh analysis to the circuit of Figures 5.9 and 5.10, we then obtain (5.11):

$$Z_{\varphi\varphi}I_\varphi = \tilde{e}_\varphi \tag{5.11}$$

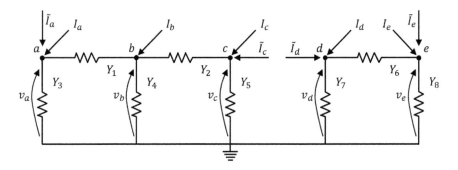

Figure 5.8 Partitioned example circuit

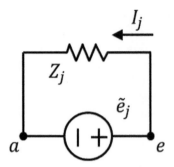

Figure 5.9   Removed branch j

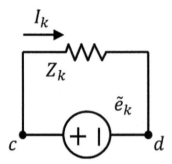

Figure 5.10   Removed branch k

where:

$$\tilde{e}_\varphi = \begin{bmatrix} \tilde{e}_j & \tilde{e}_k \end{bmatrix} \tag{5.12}$$

$$I_\varphi = \begin{bmatrix} \tilde{I}_j & \tilde{I}_k \end{bmatrix} \tag{5.13}$$

$$Z_{\varphi\varphi} = \begin{bmatrix} Z_j & 0 \\ 0 & Z_k \end{bmatrix} \tag{5.14}$$

The decoupled system with the additional constrain is so described by (5.9) and (5.11) but we cannot solve this set of equations since we have more unknowns than equations, in addition to v_a and I_φ we also have $\tilde{e}_\varphi$ and $\tilde{I}_a$. $\tilde{e}_\varphi$ and $\tilde{I}_a$ effectively represent how the separate subsystems interact with each other. Let look—with (5.15)—at the relation that exist between $\tilde{I}_a$ and I_φ:

$$\tilde{i}_a = \begin{bmatrix} \tilde{i}_a \\ \tilde{i}_b \\ \tilde{i}_c \\ \tilde{i}_d \\ \tilde{i}_e \end{bmatrix} = \begin{bmatrix} 1 & 0 \\ 0 & 0 \\ 0 & -1 \\ 0 & 1 \\ -1 & 0 \end{bmatrix} \begin{bmatrix} i_j \\ i_k \end{bmatrix} \tag{5.15}$$

And in a more general and compact form, we can write (5.16):

$$\tilde{i}_\alpha = C_{\alpha\varphi} i_\varphi \tag{5.16}$$

The matrix $C_{\alpha\varphi}$ can be easily constructed by inspection looking at the circuit of Figure 5.8.

Similarly, the relation between $\tilde{e}_\varphi$ and v_α is reported in (5.17) and (5.18):

$$\tilde{e}_\varphi = \begin{bmatrix} \tilde{e}_j \\ \tilde{e}_k \end{bmatrix} = \begin{bmatrix} -1 & 0 & 0 & 0 & 1 \\ 0 & 0 & 1 & -1 & 0 \end{bmatrix} \begin{bmatrix} v_a \\ v_b \\ v_c \\ v_d \\ v_e \end{bmatrix} \tag{5.17}$$

$$\tilde{e}_\varphi = B_{\varphi\alpha} v_\alpha \tag{5.18}$$

As can be seen looking at (5.15) and (5.17), the matrices $C_{\alpha\varphi}$ and $B_{\varphi\alpha}$ are strictly related—they represent the same partitioning—and it is possible to write:

$$B_{\varphi\alpha} = -C_{\varphi\alpha}{}^t \tag{5.19}$$

$$\tilde{e}_\varphi = B_{\varphi\alpha} v_\alpha = -C_{\varphi\alpha}{}^t v_\alpha \tag{5.20}$$

Using the just defined relations, it is possible to rewrite (5.9) and (5.11) as (5.21) and (5.22):

$$Z_{\varphi\varphi} i_\varphi = -C_{\varphi\alpha}{}^t v_\alpha \tag{5.21}$$

$$\tilde{Y}_{\alpha\alpha} v_\alpha = I_\alpha + C_{\alpha\varphi} i_\varphi \tag{5.22}$$

Or in a more compact matricial form

$$\begin{bmatrix} Z_{\varphi\varphi} & C_{\varphi\alpha}{}^t \\ -C_{\alpha\varphi} & \tilde{Y}_{\alpha\alpha} \end{bmatrix} \begin{bmatrix} i_\varphi \\ v_\alpha \end{bmatrix} = \begin{bmatrix} 0 \\ I_\alpha \end{bmatrix} \tag{5.23}$$

The system of equations in (5.23) is normally indicated as the fundamental equation of diakoptics and they could be used to calculate the value of v_α, at the same time, it is important to understand how to approach the solution of (5.23) since it potentially could lead to a more computationally expensive process that the one required by the direct solution of the original problem described by (5.1). It can be easily seen that (5.23) will always have an higher dimension than (5.1). The number of equations composing (5.23) is equal to the one of (5.1) plus a number of equation equal to the number of constrains and so or removed branches.

Combining (5.21) into (5.22), we can than write (5.24):

$$v_\alpha = \tilde{Y}_{\alpha\alpha}{}^{-1} \left(I_\alpha + C_{\alpha\varphi} i_\varphi \right) \tag{5.24}$$

Substituting (5.24) back into (5.21), we then obtain (5.25)

$$Z_{\varphi\varphi} i_\varphi = -C_{\varphi\alpha}{}^t \tilde{Y}_{\alpha\alpha}{}^{-1} \left(I_\alpha + C_{\alpha\varphi} i_\varphi \right) = -C_{\varphi\alpha}{}^t \tilde{Y}_{\alpha\alpha}{}^{-1} I_\alpha - C_{\varphi\alpha}{}^t \tilde{Y}_{\alpha\alpha}{}^{-1} C_{\alpha\varphi} i_\varphi \tag{5.25}$$

We can then manipulate (5.25) so to obtain (5.28). Intermediate steps (5.26) and (5.27) are also reported:

$$\left(Z_{\varphi\varphi} + C_{\varphi\alpha}{}^t\tilde{Y}_{\alpha\alpha}{}^{-1}C_{\alpha\varphi}\right)i_\varphi + C_{\varphi\alpha}{}^t\tilde{Y}_{\alpha\alpha}{}^{-1}I_\alpha = 0 \tag{5.26}$$

$$\tilde{Z}_{\varphi\varphi} = \left(Z_{\varphi\varphi} + C_{\varphi\alpha}{}^t\tilde{Y}_{\alpha\alpha}{}^{-1}C_{\alpha\varphi}\right) \tag{5.27}$$

$$i_\varphi = -\tilde{Z}_{\varphi\varphi}{}^{-1}C_{\varphi\alpha}{}^t\tilde{Y}_{\alpha\alpha}{}^{-1}I_\alpha \tag{5.28}$$

If we now substitute back (5.28) into (5.24), we obtain (5.29)

$$v_\alpha = \tilde{Y}_{\alpha\alpha}{}^{-1}I_\alpha - \tilde{Y}_{\alpha\alpha}{}^{-1}C_{\alpha\varphi}\tilde{Z}_{\varphi\varphi}{}^{-1}C_{\varphi\alpha}{}^t\tilde{Y}_{\alpha\alpha}{}^{-1}I_\alpha \tag{5.29}$$

To take full advantage of the block diagonal form of matrices $\tilde{Y}_{\alpha\alpha}$ and $\tilde{Z}_{\varphi\varphi}$, steps described in (5.30) through (5.37):

$$\tilde{Z}_{\alpha\alpha} = \tilde{Y}_{\alpha\alpha}{}^{-1} \tag{5.30}$$

$$Y'_{\varphi\varphi} = \tilde{Z}_{\varphi\varphi}{}^{-1} \tag{5.31}$$

$$v'_\alpha = \tilde{Z}_{\alpha\alpha}I_\alpha \tag{5.32}$$

$$v'_\varphi = C_{\varphi\alpha}{}^t v'_\alpha \tag{5.33}$$

$$i_\varphi = Y'_{\varphi\varphi}v'_\varphi \tag{5.34}$$

$$i'_\alpha = C_{\alpha\varphi}i_\varphi \tag{5.35}$$

$$v''_\alpha = \tilde{Z}_{\alpha\alpha}i'_\alpha \tag{5.36}$$

$$v_\alpha = v'_\alpha - v''_\alpha \tag{5.37}$$

We proceed now providing a couple of example of how diakoptics can be effectively used to simulate an electrical circuit.

Example 1 As the first example, we can proceed numerically solving the circuit of Figure 5.7 using the value of Table 5.1.

Table 5.1   Example 1 Parameters

Component	Value
$R_1, R_2, R_3, R_4, R_5, R_6, R_7, R_8, R_9, R_{10}$	$1\,\Omega$
I_a, I_c, I_e	$1\,A$
I_b, I_d	$0\,A$

As the first step, we create the reference solution applying nodal analysis to the unpartitioned circuit so obtaining the vector and matrix of (5.38) and (5.39):

$$I_\alpha = \begin{bmatrix} 1 & 0 & 1 & 0 & 1 \end{bmatrix}^T \tag{5.38}$$

$$Y_{\alpha\alpha} = \begin{bmatrix} 3 & -1 & 0 & 0 & -1 \\ -1 & 3 & -1 & 0 & 0 \\ 0 & -1 & 3 & -1 & 0 \\ 0 & 0 & -1 & 3 & -1 \\ -1 & 0 & 0 & -1 & 3 \end{bmatrix} \tag{5.39}$$

Solving (5.1), we obtain the node voltages:

$$v_\alpha = \begin{bmatrix} 0.73 & 0.45 & 0.64 & 0.45 & 0.73 \end{bmatrix}^T \tag{5.40}$$

With reference to (5.5), (5.6), (5.14), and (5.17), we can write the matrices that characterize the partitioned system:

$$Y_{11} = \begin{bmatrix} 2 & -1 & 0 \\ -1 & 3 & -1 \\ 0 & -1 & 2 \end{bmatrix} \tag{5.40a}$$

$$Y_{22} = \begin{bmatrix} 2 & -1 \\ -1 & 2 \end{bmatrix} \tag{5.41}$$

$$B_{\varphi\alpha} = \begin{bmatrix} -1 & 0 & 0 & 0 & 1 \\ 0 & 0 & 1 & -1 & 0 \end{bmatrix} \tag{5.42}$$

$$Z_{\varphi\varphi} = \begin{bmatrix} 1 & 0 \\ 0 & 1 \end{bmatrix} \tag{5.43}$$

Applying the solution process defined by (5.30)–(5.37), we can then calculate the node voltages, as mentioned before the solution is identical to the one computed using the unpartitioned model:

$$v_\alpha = \begin{bmatrix} 0.73 & 0.45 & 0.64 & 0.45 & 0.73 \end{bmatrix}^T \tag{5.44}$$

Example 2 As the second example, we consider a simple dynamic electrical circuit, Figure 5.11 and Table 5.2. As mentioned above, once a dynamic—even non-linear—circuit is modeled using the resistive companion approach diakoptics can be

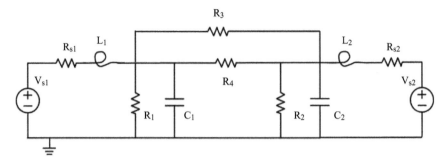

Figure 5.11   Linear dynamic circuit

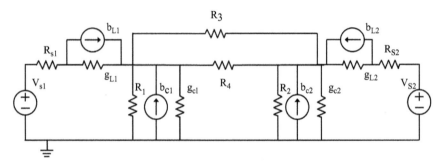

Figure 5.12   Discretized equivalent of the circuit of Figure 5.5

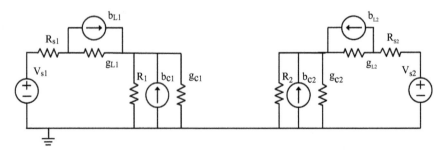

Figure 5.13   Partitioned version of the circuit of Figure 5.6

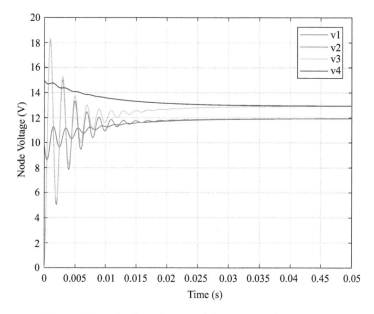

Figure 5.14 Node voltages of the circuit of Figure 5.7

Table 5.2 Example 2 parameters

V_{s1}	10 V
V_{s2}	15 V
$R_{s1}\ R_{s2}\ R_3\ R_4$	0.5 Ω
L_1	1 mH
L_2	10 mH
C_1	100 μF
C_2	10 μF
R_1	100 Ω
R_2	150 Ω
$I_L(0)$	0 A
$V_C(0)$	0 V

so applied to the obtained static equivalent circuit. Figure 5.6 shows the obtained equivalent circuit; we used trapezoidal rule for discretization and we remand the reader to Chapter 3 for reference.

Now that the dynamic circuit has been transformed in an equivalent static one it can be partitioned using Diakoptics. To create subsystems of similar size we removed R_3 and R_4. Figure 5.7 shows the obtained decoupled system. The simulation results from the decupled system are identical to the one of the original one, in Figure 5.8, we show the node voltages during the first 50 ms. Try to implement the model using Diakoptics and compare your results to the one of Figure 5.8.

5.5   State-space nodal method (SSN)

In [11], the authors propose a new solution method that combines state space and nodal representation. The two main advantages of such method are: the possibility of benefiting from the advantages of the most suitable method, which can so be freely selected, and the possibility of parallelizing execution. Another benefit of this approach is the possibility of maintaining the solution of switching devices independent. This is extremely relevant since a solution flow that considers pre calculation of state matrix has been adopted in [11]: partitioning the system and keeping the switching devices separate reduces the combinations and the number of matrices to be recalculated and stored.

The main idea of the SSN approach is to combine arbitrary groups of components—modeled using state equations—in a single nodal admittance matrix. All components of the system are discretized using trapezoidal rule. So let us start assuming that any group of components is modeled using a state-space approach as in (5.45) and (5.46):

$$\dot{x} = A_k x + B_k u \tag{5.45}$$

$$y = C_k x + D_k u \tag{5.46}$$

The subscript k indicates the kth permutation associated with each enumerated switches position. Since we assumed that each component is discretized using trapezoidal rule, discretizing (5.45) we obtain (5.47), that as expected include the input vector from two different instant of time t and $t + \Delta t$

$$x_{t+\Delta t} = \widehat{A}_k x_t + \widehat{B}_k u_t + \widehat{B}_k u_{t+\Delta t} \tag{5.47}$$

$$y_{t+\Delta t} = C_k x_{t+\Delta t} + D_k u_{t+\Delta t} \tag{5.48}$$

Equations (5.47) and (5.48) can be re-written separating internal sources indicated by the i subscript and external nodal injections indicated by the subscript n. We can then combine (5.49) and (5.50) so to obtain (5.51):

$$x_{t+\Delta t} = \widehat{A}_k x_t + \widehat{B}_k u_t + \begin{bmatrix} \widehat{B}_{ki} & \widehat{B}_{kn} \end{bmatrix} \begin{bmatrix} u^i_{t+\Delta t} \\ u^n_{t+\Delta t} \end{bmatrix} \tag{5.49}$$

$$\begin{bmatrix} y^i_{t+\Delta t} \\ y^n_{t+\Delta t} \end{bmatrix} = \begin{bmatrix} C_{ki} \\ C_{kn} \end{bmatrix} x_{t+\Delta t} + \begin{bmatrix} D_{kii} & D_{kin} \\ D_{kni} & D_{knn} \end{bmatrix} \begin{bmatrix} u^i_{t+\Delta t} \\ u^n_{t+\Delta t} \end{bmatrix} \tag{5.50}$$

$$y^n_{t+\Delta t} = C_{kn}\left(\widehat{A}_k x_{t+\Delta t} + \widehat{B}_k u_t + \widehat{B}_{ki} u^i_{t+\Delta t}\right) + D_{kni} u^i_{t+\Delta t} + \left(C_{kn}\widehat{B}_{kn} + D_{knn}\right) u^n_{t+\Delta t} \tag{5.51}$$

Equation (5.51) can be re-arranged to isolate a known term—indicated as *hist*—so to obtain (5.52), where the term W_{kn} is defined by (5.53):

$$y^n_{t+\Delta t} = y_{k\,hist} + W_{kn} u^n_{t+\Delta t} \tag{5.52}$$

$$W_{kn} = \left(C_{kn}\widehat{B}_{kn} + D_{knn}\right) \tag{5.53}$$

At this point, it is convenient to make a distinguish between I-type SSN group and V-type SSN group. We identify a V-type SSN group if y^n represent a current injection and u^n is a node voltage and a I-type SSN group if y^n represent a voltage and u^n is a current entering a group. In the first case, the W_{kn} has the dimension of an admittance and in the second case of an impedance. Since in general it is possible to have both types, we can write (5.54) to cover both cases:

$$\begin{bmatrix} v^{nI}_{t+\Delta t} \\ i^{nV}_{t+\Delta t} \end{bmatrix} = \begin{bmatrix} v_{k\,hist} \\ i_{k\,hist} \end{bmatrix} + \begin{bmatrix} W_{II} & W_{IV} \\ W_{VI} & W_{VV} \end{bmatrix} \begin{bmatrix} i^{nI}_{t+\Delta t} \\ v^{nV}_{t+\Delta t} \end{bmatrix} \tag{5.54}$$

To combine the state-space representation of (5.54) with a nodal one, it is necessary to manipulate (5.54) so to have only current term on the left side so obtaining (5.55):

$$\begin{bmatrix} i^{nI}_{t+\Delta t} \\ i^{nV}_{t+\Delta t} \end{bmatrix} = \Gamma^n_k \begin{bmatrix} v_{k\,hist} \\ i_{k\,hist} \end{bmatrix} + Y^n_k \begin{bmatrix} v^{nI}_{t+\Delta t} \\ v^{nV}_{t+\Delta t} \end{bmatrix} \tag{5.55}$$

We can finally reconnect the state-space solution with the global nodal admittance by mapping the matrix Y^n_k node with the one of the nodal representations of (5.56):

$$i^N_{t+\Delta t} = Y^N v^N_{t+\Delta t} \tag{5.56}$$

According to the SSN approach, the solution process can be divided in steps as indicated below:

Step (1): Find the steady-state solution.

Step (2): Advance to the next time-point.

Step (3): Determine all switch positions (th permutation) and formulate state-space (5.45) and (5.50).

Step (4): Determine all history terms and update.

Step (5): Update (if necessary) the global nodal admittance matrix that contains contributions from all groups.

Step (6): Update i^N from group contributions.

Step (7): Solve the nodal system to determine all node voltages.

Step (8): Use (5.45) and (5.50) to compute the state-space solutions at the current time-point.

Step (9): Go back to Step (2) if the simulation did not reach the last time-point.

5.6 Transmission line modeling and the waveform relaxation-based method

Several methods have been developed that exploiting this natural property parallelize simulation execution. In this chapter, we will focus on methods that have been defined in the recent years mainly to support real-time execution but before doing that we will provide an overview on the two historically most significant simulation methods that exploit latency to parallelize simulation execution: the transmission line modeling (TLM) and the waveform relaxation based methods.

TLM has been first introduced as a time-domain, differential numerical technique for simulating field problems in 1971 by Johns and Beurle [15], lately in 1980, Johns and O'Brien [16] applied TLM to the simulation of lumped parameter nonlinear network. For a review of the first development of TLM, refer to [17]. In 1980, TLM was used for the first time to simulate power electronics circuits. In [18], Hui and Christopoulos extended the method to include the modeling of switching devices expanding the field of use of TLM to power converters simulation. The main idea of [18] is to replace all circuit components with transmission line segment (stubs). The propagation time on each stub is the same for the whole network and it is equal to the simulation time step. By proper tuning of the stubs, it is possible to give to each component a predominant inductive or capacitive behavior, respectively, capacitive and inductive parasitic effect cannot be reduced to zero and represent the main source of error in the proposed modeling approach. Switches can be modeled as capacitance or inductive stubs depending on their state. The modeling error introduced by the parasitic inductance and capacitance can be limited reducing the time step. If accurate results are required, the time step has to be chosen in relation to the smallest value of pure capacitance or inductance that has to be modeled. When applied to simulation of a power converter, this may lead to the use of very small time steps. In [18], a time step of 60 ns was selected in front of a switching frequency of 27 kHz. At the same time, this method has very good property of stability and it is computationally extremely efficient, a solution to allow multi-rating execution of the TLM is presented in [19] solving the inconvenience related to the requirement of a single time step for the simulation of the overall network.

Applying to the simulation of power electronics circuit also the concept of links TLM has been used in [20,21] to decouple the network and to allow simulation parallelization. Using links to decouple, the system introduces an error proportional to the one time step delay that exist in the exchange of information between two ends of a TLM model.

A further extension of nonlinear TLM was presented in [22]. In [23], a resistive companion model of a link was introduced so that TLM could be used to parallelize nodal analysis.

Another important family of solution methods that leverage on latency exploitation to parallelize and speed up electric transient simulation is the on one of the waveform relaxation based methods [24–26]. According to the waveform relaxation methods, a system of DAE if solved using Gauss-Seidel and Gauss-Jacobi relaxation methods, originally defined for the solution of large system of algebraic equations.

According to the waveform relaxation method, a system of DAE is separate in subsystems that are solved independently at each iteration. Each subsystem uses the waveform values calculated at the previous iteration as guess of the state of the other subsystems. At each iteration, information are exchanged between the subsystems.

The waveform relaxation method has been used successfully for special type of circuits but at the same time for many other types of circuit, the performance, in terms of accuracy and stability, are really poor [27]. If compared to direct solution by implicit integration of the system of differential equations, iterative matrices methods present slow rates of convergence and are less robust.

At the same time, one of the main advantages presented by the waveform relaxation algorithm is the high level of parallelizability offered. Every subsystem can be assigned to a different processor/kernel achieving significant speed up; at the same time, the need of multiple iterations during the same time step make this system unfeasible for real-time execution. The waveform relation algorithm was used for the first for the solution of power system transient simulation in [28] and in [29] the parallel processing possibilities are investigated.

5.7 Latency insertion method

Latency insertion method (LIM), first introduced in [30], is a finite difference method for transient circuit simulation. Let us assume a circuit with reactive latency in all branches and nodes so that branch currents are continuous first-order functions of time, and node voltages are continuous first-order functions of time. Then, as defined by the LIM, it is possible to solve the networks through a set of algebraic steps. LIM has linear computational complexity and moreover, each algebraic equation can be potentially solved in parallel, ensuring perfect scalability.

A network has latency if each branch of the network contains an inductance and each node of the network provides a capacitive path to ground; if these values are not naturally present or if the present values are small and thus present latency much smaller than the time features of interest, additional capacitance or inductance can be added to increase latency.

As shown in Figure 5.15(a), any generic branch is composed of a series combination of a resistor, an inductor and a voltage source; applying KVL to the circuit of Figure 5.15(a), we can write the characteristic branch equation as:

$$V_i^{n+\frac{1}{2}} - V_j^{n+\frac{1}{2}} = L_{ij}\left(\frac{I_{ij}^{n+1} - I_{ij}^n}{\Delta t}\right) + R_{ij}I_{ij}^n - E_{ij}^{n+\frac{1}{2}} \tag{5.57}$$

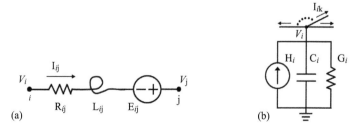

Figure 5.15 (a) Generic LIM branch and (b) generic LIM node

From (5.57), it is possible to calculate the unknown current:

$$I_{ij}^{n+1} = I_{ij}^n + \frac{\Delta t}{L_{ij}}\left(V_i^{n+\frac{1}{2}} - V_j^{n+\frac{1}{2}} - R_{ij}I_{ij}^n + E_{ij}^{n+\frac{1}{2}}\right) \tag{5.58}$$

Equation (5.58) must be computed for all the branches of the network at each time step.

As shown in Figure 5.15(b), each generic node is composed of a parallel combination of a conductance, a capacitor and a current source; applying KCL to the circuit of Figure 5.15(b), we can write the characteristic node equation as:

$$C_i\left(\frac{V_i^{n+\frac{1}{2}} - V_i^{n-\frac{1}{2}}}{\Delta t}\right) + G_i V_i^{n+\frac{1}{2}} - H_i^n = -\sum_{j=1}^{M_i} I_{ik}^n \tag{5.59}$$

where M_i is the number of branches connected to the node i. From (5.59), it is possible to calculate the unknown voltage:

$$V_i^{n+\frac{1}{2}} = \frac{\frac{C_i V_i^{n-\frac{1}{2}}}{\Delta t} + H_i^n - \sum_{j=1}^{M_i} I_{ik}^n}{\frac{C_i}{\Delta t} + G_i} \tag{5.60}$$

Equation (5.60) is computed for all the nodes of the network at each half time step.

Using a leapfrog approach, the current through each branch and the voltage at each node can be updated alternately. The time is discretized and the current and voltage quantities are allocated in half time step; see Figure 5.16.

As can be noticed by (5.58) and (5.60), the LIM applies a semi-implicit leap-frog integration method. As a consequence, the stability of the method is a concern as the choice of time step for stable operation is conditional to the inductance and capacitance found in a given model under LIM. The stability of basic LIM and of its variations has been analyzed in many papers [30–37]. In [30], the stability of LIM using leap-frog integration is determined using Telegrapher's Equations for a per-cell basis (one branch and one node connected together in a model), the stability condition being

$$\Delta t < \sqrt{L_k C_k}, \tag{5.61}$$

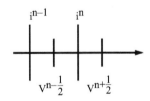

Figure 5.16   Time line structure

where k is the index of the cell with the smallest latency derived from L_k and C_k in a given model. In [31,32], further detailed stability analysis of LIM was explored for RLC and GLC topologies using Lyapunov's stability theorem, though the general case of RLCG topologies was not studied. The stability analysis of the use of explicit, semi-implicit, and implicit integration methods in LIM was presented in [33]. The stability of LIM was analyzed in [34] using block-matrix formulation, determining the stability of a system for a given time step using the eigenvalues of a computed amplification matrix for said system. A general stability criteria is given in [35] for partitioned LIM which highlights stability for models where branches and nodes contain dependent sources through which components are interconnected. In [35], an approach is proposed and analyzed for improving stability of LIM to be unconditional to latency or time step through introducing a single formulation to update branches and nodes, performed through implicitly substituting branch currents and node voltages into said formulation.

Example 3 To better illustrate the use of LIM for the solution of an electrical network, let us consider the circuit of Figure 5.17, the circuit it is already separated in two LIM compatible nodes and one branch. The parameter of the circuit is reported in Table 5.3.

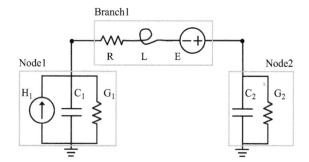

Figure 5.17 LIM example circuit

Table 5.3 Example 3 parameters

H_1	1 A
C_1	0.1 mF
G_1	1 mS
L	10 mH
R	1 Ω
E	2 V
C_2	0.1 mF
G_2	0.1 S
$I_L(0)$	0 A
$V_{C1}(0)$, $V_{C2}(0)$	0 V

Equations (5.62)–(5.64)—that fully describe the circuit of interest—are obtained applying the (5.58) and (5.60) to the specific circuit of Figure 5.17.

In Figure 5.18, we report the simulation results obtained using the LIM for the calculation of the branch current (inductor current) and nodes voltage (capacitors voltage). A detail of the first instant of simulation is reported in Figure 5.19, it can be noticed how the update of the voltage and the current is shifted of half time step as indicated by the leapfrog approach.

$$I_L^{n+1} = I_L^n + \frac{\Delta t}{L}\left(V_{C1}^{n+\frac{1}{2}} - V_{C2}^{n+\frac{1}{2}} - RI_L^n + E^{n+\frac{1}{2}}\right) \tag{5.62}$$

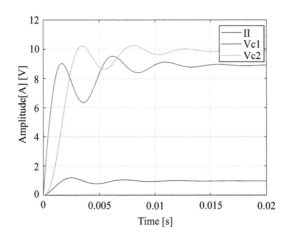

Figure 5.18   LIM example simulation results

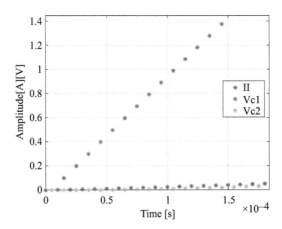

Figure 5.19   Detail of LIM example simulation results

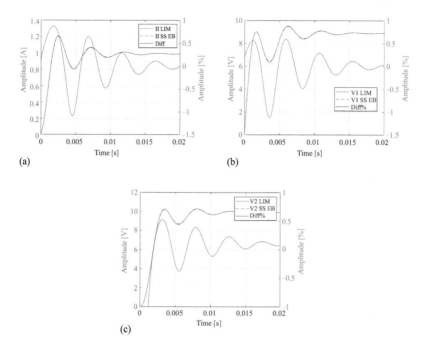

Figure 5.20 *Comparison between LIM and SS EB solution: (a) branch current,*
(b) node 1 voltage, and (c) node 2 voltage

$$V_{C1}^{n+\frac{1}{2}} = \frac{\frac{C_1 V_{C1}^{n-\frac{1}{2}}}{\Delta t} + H_1^n - I_L^n}{\frac{C_1}{\Delta t} + G_1} \tag{5.63}$$

$$V_{C2}^{n+\frac{1}{2}} = \frac{\frac{C_2 V_{C2}^{n-\frac{1}{2}}}{\Delta t} + I_L^n}{\frac{C_2}{\Delta t} + G_2} \tag{5.64}$$

For reference in Figure 5.20, we compare the solution obtained with LIM and
the one obtained from the solution of a state-space (SS) representation of the circuit
solved using Euler backward (EB). Since we are operating with a time step that
ensure a stable integration of the circuit under test, the accuracy of the two methods
is comparable and the difference between the two solutions is neglectable.

5.7.1 Latency insertion method for power electronics simulation

Despite the conditional stability of leapfrog integration in LIM, this integration
structure provides benefits to modeling high frequency switching power electronic
converters. This semi-implicit integration structure separates branch current and
node voltage updates into computationally cheap explicit difference half time step

terms (similar to Euler forward) which can be exploited to model converter switching behavior in a single time step without the need to use a fully implicit approach on entire converter model; while still maintaining second order numerical integration accuracy provided by leapfrog integration. By having the current and voltage terms integrated separately, this structure allows representing ideal switching phenomena by linking nodes and branches through their respective ideal voltage and current sources (E, H), without introducing any additional delay in the integration; Moreover, this linking maintains full compatibility with the LIM formulation and stability analysis presented in [26], where voltage and current sources can depend on other quantities in the system.

As the LIM was originally developed for the simulation of semiconductor passive transmissions lines which consist of inductance with series resistance and capacitance with parallel conductance, limitations exist in adapting the method for power electronic systems. Elements such as branch capacitors or node inductors which are often found in some power electronic systems cannot be directly modeled using traditional LIM without deviating from the ideal modeling of such elements, often requiring said components to be modeled as their integration companion model equivalents that incorporate inserted fictitious latency to fit LIM structure. Moreover, some topologies, such as wye or delta arrangement of inductors seen in three-phase converter systems, cannot be modeled in LIM without inserting fictitious, small capacitances to ground at each node connection of the inductor branches. If a converter topology contains floating branch or node elements, techniques beyond traditional LIM would be needed to model mutual capacitance or inductance between such elements, such as seen in [37]. Along with topology restrictions, the LIM expects sufficient latency to exist in all branch and node elements of a given network topology while still achieving feasible and stable time steps; should a system incorporate small latency elements, additional latency is to be inserted, which is undesirable as doing so will deviate the model from the original system. Despite these limitations, sufficient number of common power electronic systems can be effectively modeled in LIM without deviating significantly, or at all, from the overall characteristics of the system in question. In this section of the chapter, we review how a buck converter can be modeled using the LIM, other topologies are analyzed in [38].

Let us take as a reference the ideal buck converter shown in Figure 5.21. In the buck converter, the output filtering capacitor and load resistance are mappable to a

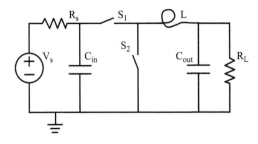

Figure 5.21   Buck converter

LIM node, the inductor is mappable to a LIM branch, and the voltage source input and input capacitor are mappable to another LIM node after the voltage source has been transformed into a current source with Norton's transformation. The nonlinear switching elements of the converter are left out of the LIM topology. This mapping of the converter to LIM components is shown in Figure 5.22.

The question arises on how to handle the nonlinear switching action of the buck converter with LIM without incorporating the switching elements in the LIM topology. As seen in previous discussion on LIM, branch and node models contain, respectively, a voltage source E and a current source H which can be arbitrarily altered during simulation. Using these model sources, the switching action can be handled by altering the values of H and E in LIM component models in sync with the switching states of this simulated converter, while keeping the LIM topology fixed regardless of the switching state; this is unlike typical state equation modeling of converters where the state model of the converter is changed within a state machine for each switching topology the converter is in. Let us first consider the continuous conduction mode using traditional power electronic system analysis.

As shown in Figure 5.23(a) and (b) during switch-on state, the E term of the inductor branch is equated to the input voltage across the input capacitor node, and the H term of the input capacitor node is equated to the inductor branch current.

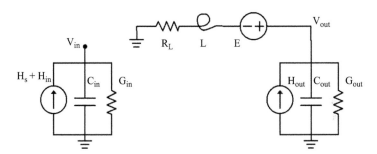

Figure 5.22   Buck converter LIM model

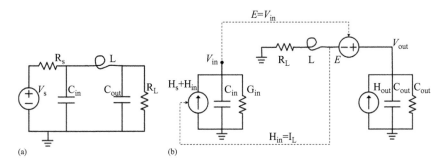

Figure 5.23   (a) Buck converter in continuous conduction mode on-state and (b) equivalent LIM model

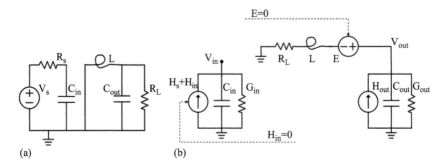

Figure 5.24 *(a) Buck converter in continuous conduction mode off-state and (b) equivalent LIM model*

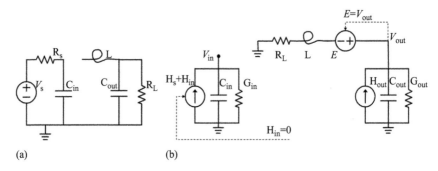

Figure 5.25 *(a) Buck converter in discontinuous conduction mode off-state and (b) equivalent LIM model*

During the off state of this converter, both H and E are set to zero, as in Figure 5.24 (a) and (b). By altering H and E terms according to a switching control signal in this fashion, the switching action of the converter using LIM is simulated.

To also consider discontinuous conduction mode off-state Figure 5.25(a), the model monitors the current through the inductance during the off state; if the current crosses zero, the E term of the inductor branch equates to the output voltage, displayed in Figure 5.25(b).

5.7.2   *Latency insertion method combined with state space and nodal methods*

While in many cases, electrical circuits can be modeled using the topology required by LIM, in same case, this is simply impossible or inconvenient—may be because a large piece of the system has already been modeled using a more convenient method. To still take partial advantage of the LIM parallel execution characteristics in [39], an approach to combine LIM with state space and nodal methods are presented. Effectively in the approach proposed in [39] LIM is mainly used to represent the

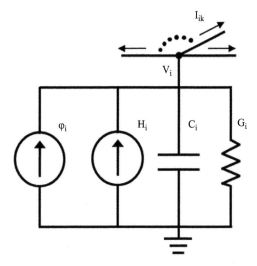

Figure 5.26   General LIM node model

interconnections of a partitioned system, following this approach each sub-system can be solved in parallel and the solution of the LIM part of the system ensure that boundary constrain is imposed. To do this, a slightly modified node topology— Figure 5.26—is introduced. H_i represents the net branch current as computed internal to the LIM solver, and φ_i represents the net current injection from the subsystem that is computed external to the LIM solver. Since we are modifying the traditional LIM topology and introducing a time step delay between the LIM and the subsystem solution, it is important to study the stability characteristics of this approach.

References [34,35] described a block matrix formulation for LIM and used that to analyze stability for the more general RLCG case and they also eliminated the simplifying assumptions related to current and voltage independence. But still, in [34,35], the complete network was solved using LIM, and so it is clear that voltage and current insertions may depend only on other voltages or currents that are internal to the LIM solution. So we cannot straightforwardly use the method described in [34,35] to analyze the stability of the modified LIM, but still we will start from that point.

From now on we assume that the sub-system solution is computed at time n which also corresponds to the time that current in the branches is computed via (5.58) using LIM. The zonal solution is assumed to be computed with a nodal or a state space based solver where both V and I are computed at the same time instant n. It is assumed that both the sub-system solutions as well as the LIM network solution step at the same times.

For the stability study, we assume that current is solved first at each full time step. Let us assume that a sub-system connected to node i is already discretized and represented by (5.65), where u_i is a vector of inputs that contains all the sources (both current sources and voltage sources), internal to the sub-system and a voltage source V_i that represents the node voltage found from the LIM network solution. This voltage is updated at every time step, at the half intervals, by the LIM solver:

$$x_i^n = A_i x_i^{n-1} + B_i u_i^n \tag{5.65}$$

Equation (5.65) can be augmented with an output equation like that shown in (5.66) where the only output is the current absorbed by the voltage source V_i:

$$\varphi_i^n = C_i x_i^n + D_i u_i^n \tag{5.66}$$

Expanding the input vector, (5.65) and (5.66) can be represented in the form of (5.67) and (5.68):

$$x_i^n = A_i x_i^{n-1} + \begin{bmatrix} B_{i1} & B_{i2} \end{bmatrix} \begin{bmatrix} V_i^{n-\frac{1}{2}} \\ u_i^n \end{bmatrix} \tag{5.67}$$

$$\varphi_i^n = C_i x_i^n + \begin{bmatrix} d_{i1} & D_{i2} \end{bmatrix} \begin{bmatrix} V_i^{n-\frac{1}{2}} \\ u_i^n \end{bmatrix} \tag{5.68}$$

B_{i1} has size $(p_i \times 1)$.
B_{i2} has size $(p_i \times q_i)$.
d_{i1} has size (1×1).
D_{i2} has size $(1 \times q_i)$.
p_i is the number of internal variables of subsystem i.
q_i is the number of internal inputs of subsystem i.

Equations (5.67) and (5.68) of every sub-system can be organized in a single matrix formulation as shown in (5.69) and (5.70):

$$\begin{bmatrix} x_1^n \\ x_2^n \\ \vdots \\ x_k^n \end{bmatrix} = \begin{bmatrix} A_1 & & & \\ & A_2 & & \\ & & \ddots & \\ & & & A_k \end{bmatrix} \begin{bmatrix} x_1^{n-1} \\ x_2^{n-1} \\ \vdots \\ x_k^{n-1} \end{bmatrix}$$

$$+ \begin{bmatrix} B_{11} & & & B_{12} & & \\ & B_{21} & & & B_{22} & \\ & & \ddots & & & \ddots \\ & & & B_{k1} & & & B_{k2} \end{bmatrix} \begin{bmatrix} V_i^{n-\frac{1}{2}} \\ V_i^{n-\frac{1}{2}} \\ \vdots \\ V_i^{n-\frac{1}{2}} \\ u_1^n \\ u_2^n \\ \vdots \\ u_k^n \end{bmatrix}$$

$$\tag{5.69}$$

$$\begin{bmatrix} \varphi_1^n \\ \varphi_2^n \\ \vdots \\ \varphi_k^n \end{bmatrix} = \begin{bmatrix} C_1 & & & \\ & C_2 & & \\ & & \ddots & \\ & & & C_k \end{bmatrix} \begin{bmatrix} x_1^n \\ x_2^n \\ \vdots \\ x_k^n \end{bmatrix}$$

$$+ \begin{bmatrix} d_{11} & & & & D_{12} & & \\ & d_{21} & & & & D_{22} & \\ & & \ddots & & & & \ddots & \\ & & & d_{k1} & & & & D_{k2} \end{bmatrix} \begin{bmatrix} v_i^{n-\frac{1}{2}} \\ v_i^{n-\frac{1}{2}} \\ \vdots \\ v_i^{n-\frac{1}{2}} \\ u_1^n \\ u_2^n \\ \vdots \\ u_k^n \end{bmatrix} \quad (5.70)$$

These can be written more compactly as (5.71) and (5.72):

$$x^n = Ax^{n-1} + [B_1 \quad B_2] \begin{bmatrix} v^{n-\frac{1}{2}} \\ u^n \end{bmatrix} \quad (5.71)$$

$$\varphi^n = Cx^n + [D_1 \quad D_2] \begin{bmatrix} v^{n-\frac{1}{2}} \\ u^n \end{bmatrix} \quad (5.72)$$

Substituting (5.71) into (5.72), the output equation for the subsystem can be expressed as (5.73):

$$\varphi^n = CAx^{n-1} + (CB_1 + D_1)V^{n-\frac{1}{2}} + (CB_2 + D_2)u^n \quad (5.73)$$

For the part of the network solved using LIM, an incidence matrix $M_{nn,nb}$ can be defined, where a nn is the number of nodes and nb is the number of branches.

$M_{q,p} = 1$ if branch p is incident at node q and the current flows away from node q.

$M_{q,p} = -1$ if branch p is incident at node q and the current flows into node q.

$M_{q,p} = 0$ if branch p is not incident at node q.

Using the incidence matrix M, the LIM branch (5.58) and node (5.60) equations can be reformulated in a matrix form following the approach presented in [34] so obtaining the (5.74) and (5.75):

$$i^n = Q_+Q_-i^{n-1} + Q_+M^Tv^{n-\frac{1}{2}} + Q_+e^{n-\frac{1}{2}} \quad (5.74)$$

$$v^{n+\frac{1}{2}} = P_+P_-v^{n-\frac{1}{2}} - P_+Mi^n + P_+h^n + P_+\varphi^n \quad (5.75)$$

where:

$$P_+ = \left(\frac{C}{t} + \frac{G}{2} \right)^{-1} \quad (5.76)$$

$$P_- = \left(\frac{C}{t} - \frac{G}{2}\right)$$ (5.77)

$$Q_+ = \left(\frac{L}{t} + \frac{R}{2}\right)^{-1}$$ (5.78)

$$Q_- = \left(\frac{L}{t} - \frac{R}{2}\right)$$ (5.79)

After substituting (5.73) and (5.74) into (5.75), a block matrix formulation for modified LIM can be obtained by combining (5.71), (5.74), and (5.75):

$$\begin{bmatrix} v^{n+\frac{1}{2}} \\ i^n \\ x^n \end{bmatrix} = A_{\text{full}} \begin{bmatrix} v^{n-\frac{1}{2}} \\ i^{n-1} \\ x^{n-1} \end{bmatrix} + B_{\text{full}} \begin{bmatrix} h^n \\ e^{n-\frac{1}{2}} \\ u^n \end{bmatrix}$$ (5.80)

$$A_{\text{full}} = \begin{bmatrix} P_+P_- - P_+MQ_+M^T + P_+CB_1 + P_+D_1 & -P_+MQ_+Q_- & P_+CA \\ Q_+M^T & Q_+Q_- & 0 \\ B_1 & 0 & A \end{bmatrix}$$ (5.81)

$$B_{\text{full}} = \begin{bmatrix} P_+ & -P_+MQ_+ & P_+CB_2 + P_+D_2 \\ 0 & Q_+ & 0 \\ 0 & 0 & B_2 \end{bmatrix}$$ (5.82)

Equation (5.80) represents a discrete linear time invariant (DLTI) system. Accordingly, the solution will be asymptotically stable if all eigenvalues of A_{full} have magnitude strictly smaller than one.

The stability analysis has been performed assuming that the different sub-systems have been modeled in a state-space framework, a similar work could have been done assuming sub-systems have been modeled using nodal analysis. Moreover, for what concern stability analysis, we can assume that any LTI system can always be described as in (5.65) and (5.66) even if the subsystems might actually be solved using a different approach. In other words, a state space model can always be created, using the same discretization algorithm, with the purpose of studying the stability of the considered simulation scenario, even if we do not actually use the state space model to obtain the sub-systems solution.

5.8   LB-LMC method

The LB-LMC method has been defined in [9,40] with the objective of supporting multi-physics, power electronics dense, power systems real-time simulation. Naturally the solution method should be highly parallelizable, computationally as light as possible and satisfy the strict requirement of real-time execution without any risk of overrun. Another important requirement for the simulation method is that parallelization should be completely automated so not to require any specific action by the user without assuming any specific node or branch topology. The simulation method should also allow for multi-rating execution.

At the same time, the application considered presents some peculiarities that have been taken in consideration in the development of this new method. Power electronics systems simulation, if switching models are considered, is well known to require small time step size as a consequence of the relative high switching frequency used. If this requirement on one side represents a significant challenge for real-time execution, on the other side, the small time step significantly increase the latency effect of the circuit reactive elements. As a consequence, the main idea in this method is to exploit the latency directly at component level, decoupling the integration of individual components from the system integration, and thus significantly increasing the possibility to parallelize. Since the method combines state space and nodal representation of the system of interest the LB-LMC method has significant similarities with [11].

Let us start by considering a circuit composed of linear components only, except for components i and j. First assumption for applying the presented method is that every nonlinear component can be expressed as in (5.83) or (5.84). From now on, we will refer to components like (5.83) as current-type components and to components like (5.84) as voltage-type components. In general, multi-terminal components can be described as a mix of current-type and voltage-type terminal:

$$\frac{di_i^n}{dt} = f\left(v, i, x_i^n, u_i^n, t\right) \tag{5.83}$$

$$\frac{dv_j^n}{dt} = f\left(v, i, x_j^n, u_j^n, t\right) \tag{5.84}$$

where:

v is the vector of the network node voltages; i is the vector of the network branch currents; x_i^n is the vector of the state variable internal to the ith nonlinear component; u_i is the vector of the input internal to the ith the nonlinear component.

The two (5.83) and (5.84) are explicit discretized so that (5.85) and (5.86) are obtained:

$$I_i^n(k+1) = f\left(v(k), i(k), x_i^n(k), u_i^n(k), k\right) \tag{5.85}$$

$$V_j^n(k+1) = f\left(v(k), i(k), x_j^n(k), u_j^n(k), k\right) \tag{5.86}$$

Assuming more nonlinear components in the circuit and grouping them according to their type (voltage or current), (5.87) and (5.88) are obtained:

$$I^n(k+1) = f(v(k), i(k), x(k), u(k), k) \tag{5.87}$$

$$V^n(k+1) = f(v(k), i(k), x(k), u(k), k) \tag{5.88}$$

Equations (5.87) and (5.88) represent the internal step for all nonlinear components. Since the calculation of new values of voltage and current to update the relative sources only require the knowledge of the network solution at previous time step, the internal step can be easily parallelized simply broadcasting vector v and i to the designated kernels. Since the equations describing each nonlinear component are independent from one another, each kernel will solve the internal step only for a subset of the nonlinear components described by (5.87) and (5.88). The size of the subset to be solved by each kernel has to be determined on the base of the computational cost of each component and of the imposed time step size (if real-time simulation is the target). With the updated values $I^n(k+1)$ and $V^n(k+1)$, a new network solution can be computed solving the nodal equation in (5.89):

$$Gx(k+1) = b(v(k), i(k), I^n(k), V^n(k), k) \qquad (5.89)$$

Since the conductance matrix G is never updated, the LU factorization, as in linear resistive companion, can be performed off line and only the forward and backward substitutions are executed online at each iterations.

We discuss here the stability of the presented method. For the purpose of the stability analysis, we assume that both the nonlinear components and the network are modeled in a state space framework, contrary to what the presented method defines: the network should be modeled using resistive companion. The state space model of the network is formulated in (5.90). We assume that the network equations have already been discretized:

$$x(k+1) = Ax(k) + B'u'(k+1) \qquad (5.90)$$

$u'(k+1)$ is the input vector and can be separated in two parts: the vector of input sources originally connected to the network $u(k)$ and a vector of new input sources $\varphi(k)$ that represents the components solved using state equations:

$$u'(k) = \begin{bmatrix} u(k) \\ \varphi(k) \end{bmatrix} \qquad (5.91)$$

Inserting (5.91) in (5.90) and expanding, we obtain (5.92):

$$x(k+1) = Ax(k) + Bu(k+1) + B_\varphi\varphi(k+1) \qquad (5.92)$$

An output equation has to be written for the dual quantity of nonlinear component. That is, an output equation that returns the voltage at the terminals of the component has to be written for each current-type component. And vice versa, an output equation that returns the currents at the terminals of the each voltage-type component has to be written:

$$z(k+1) = Cx(k+1) + Du(k+1) + D_\varphi\varphi(k+1) \qquad (5.93)$$

In contrast to what have presented until now, here it is assumed that the components isolated from the network are also linear and can be expressed in the form of (5.94). This is done because the main purpose of the analysis presented in this paragraph is to provide a tool to easily analyze how the selection of implicit and explicit integration method affects the stability of the solution. A stability study that also considers the nonlinear case was presented in [41], the limit of this study is that allow only defining sufficient but not necessary conditions for the stability.

$$x_n(k+1) = A_n x_n(k) + B' u'_n(k) \tag{5.94}$$

Also in this case the input vector has to be separated in two parts: the vector of input sources internally connected to the network $u_n(k)$ and the new input vector $z(k)$. It is important to underline that since the components are assumed to be solved with an explicit integration method, also inputs are from the previous time step. Actually for u_n the use of the value at $(k+1)$ would not make any difference in the derivation of the following equations. It is vice versa extremely important that the value of z refers to time step (k). In this way, as said previously, the solution of the components can be computed in a single step, without iterations, starting from the network solution computed at the previous time step:

$$u'_n(k) = \begin{bmatrix} u_n(k) \\ z(k) \end{bmatrix} \tag{5.95}$$

Substituting (5.95) in (5.94), (5.96) is obtained:

$$x_n(k+1) = A_n x_n(k) + B_n u_n(k) + B_z z(k) \tag{5.96}$$

For each component, an output equation like in (5.97) has to be written. The quantity φ is the value of the equivalent source added in substitution of the component in the network solution:

$$\varphi(k+1) = C_n x_n(k+1) \tag{5.97}$$

The quantity φ has to be expressed as linear combination of state variables. Substituting (5.96) and (5.97) in (5.92), (5.98) is obtained:

$$x(k+1) = Ax(k) + Bu(k+1) + B_\varphi C_n A_n x_n(k) + B_\varphi C_n B_n u_n(k)$$
$$+ B_\varphi C_n B_z z(k) \tag{5.98}$$

Substituting now z with (5.93) from the previous time step, (5.99) is obtained:

$$x(k+1) = Ax(k) + Bu(k+1) + B_\varphi C_n A_n x_n(k) + B_\varphi C_n B_n u_n(k)$$
$$+ B_\varphi C_n B_z Cx(k) + B_\varphi C_n B_z Du(k) + B_\varphi C_n B_z D_\varphi C_n x_n(k) \tag{5.99}$$

Looking at the components equations, substituting (5.93) from the previous time step in (5.96), (5.100) is obtained:

$$x_n(k+1) = A_n x_n(k) + B_n u_n(k) + B_z Cx(k) + B_z Du(k) + B_z D_\varphi C_n x_n(k)$$
(5.100)

From (5.99) and (5.100), the overall system representation is obtained (5.101):

$$\begin{bmatrix} x(k+1) \\ x_n(k+1) \end{bmatrix} = A_{full} \begin{bmatrix} x(k) \\ x_n(k) \end{bmatrix} + B_{full1} u(k+1) + B_{full2} \begin{bmatrix} u(k) \\ u_n(k) \end{bmatrix}$$
(5.101)

where:

$$A_{full} = \begin{bmatrix} A + B_\varphi C_n B_z C & B_\varphi C_n A_n + B_\varphi C_n B_z D_\varphi C_n \\ B_z C & A_n + B_z D_\varphi C_n \end{bmatrix}$$
(5.102)

$$B_{full1} = \begin{bmatrix} B \\ 0 \end{bmatrix}$$
(5.103)

$$B_{full2} = \begin{bmatrix} B_\varphi C_n B_z D & B_\varphi C_n B_n \\ B_z D & B_n \end{bmatrix}$$
(5.104)

The solution performed with the presented LB-LMC method will be asymp-totically stable if all eigenvalues of A_{full} have magnitude strictly smaller than one. The stability of the LB-LMC has been deeply studied in [40], the LB_LMC method demonstrates to have very good stability properties and in all practical non-linear case considered the selection of the time step has always been driven by the switching frequency considered and not by the stability limit.

5.9   Exercises

Exercise 1 Let us consider the distribution network in figure and the values reported in the below table:

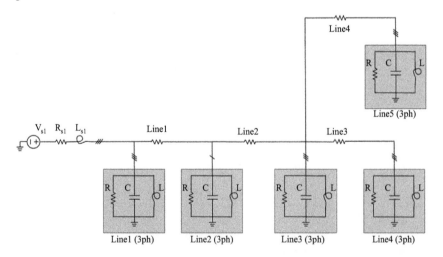

Line4

Line5 (3ph)

V_{s1} R_{s1} L_{s1} Line1   Line2   Line3

Line1 (3ph)   Line2 (3ph)   Line3 (3ph)   Line4 (3ph)

Table 5.4   Exercise 1 parameters

V_{s1} (phase to phase)	12 kV, 60 Hz
R_{s1}	0.05 Ω
L_{s1}	10 µH
R_{load}	100 kW
L_{load}	10 kVAR
C_{load}	8 kVAR
R_{line}	0.05 Ω

1. Write the conductance matrix and the source vector to solve the system using resistive companion (use nodal analysis and trapezoidal rule).
2. Calculate the value for the RLC loads assuming the whole system at nominal voltage. The values of power in table one are indicated per phase and you can use the same value for the single-phase load.
3. Simulate the system in MATLAB®. Select an appropriate time step and final time. Plot the voltage across and the current through each load.
4. Applying the concept of Diakoptics divide the system in two parts. Select where you would like to split the system between Line 1, Line 2, Line 3, and Line 4. Justify your decision.
5. Write the conductance matrix and the source vector for each of the subsystems.
6. Solve the system in MATLAB.

Exercise 2 Let us assume that you want to simulate the circuit of Figure 5.27 and that you want to separate the solution of the DC/AC converter and of the load from the one of the rest of the circuit. Let us also assume that you want to represent the DC/AC converter and the load with a state space model and the rest of the system using Resistive Companion:

1. Derive the state space model of the DC/AC converter and of the load according to the schematic of Figure 5.28 and the switching positions indicated in Table 5.2.

Table 5.5   Exercise 2 parameters

V_{grid}	100 V
R_{grid}	0.2 Ω
R_{cable}	0.3 Ω
L_{cable}	10 µH
C	10 mF
R_{load}	10 Ω
L_{load}	10 mH
f_{sw}	5 kHz
R_S	0.1 Ω
$I_L(0)$	0 A
$V_C(0)$	0 V

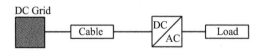

Figure 5.27   Exercise 2 single line diagram

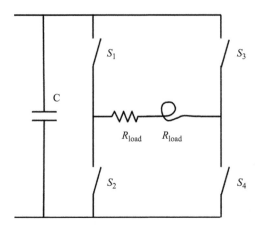

Figure 5.28   Exercise 2 converter and load schematic

2. Write the conductance matrix and the source vector of the equivalent circuit obtained, using nodal analysis and trapezoidal rule for the rest of the system (you need to derive two different systems for questions 3 and 4).
3. Simulate the system in MATLAB (select an appropriate time step and final time). Using the method described in the paper: C. Dufour, J. Mahseredjian, J. Bélanger, "A combined state-space nodal method for the simulation of power system transients", IEEE Transaction on Power Delivery, 2011: 26, no. 2.
4. Simulate the system in MATLAB (select an appropriate time step and final time). Using the method described in the paper: Benigni, A. Monti, "A parallel approach to real-time simulation of power electronics systems", IEEE Transactions on Power Electronics, 2014.

References

[1] J. Wu, A. Bose, J. Huang, A. Valette and F. Lafrance, "Parallel implementation of power system transient stability analysis", *IEEE Transactions on Power Systems*, 1995: 10, pp. 1226–1233.
[2] M. La Scala and A. Bose, "Relaxation/Newton methods for concurrent time step solution of differential-algebraic equations in power system dynamic simulations", *IEEE Transactions on Circuits and Systems I: Fundamental Theory and Applications*, 1993: 40, pp. 317–330.

[3] P. Aristidou, D. Fabozzi and T. Van Cutsem, "Dynamic simulation of large-scale power systems using a parallel Schur complement-based decomposition method", *IEEE Transactions on Parallel and Distributed Systems*, 2014: 25, no. 10, pp. 2561–2570.

[4] V. Jalili-Marandi, Z. Zhou and V. Dinavahi, "Large-scale transient stability simulation of electrical power systems on parallel GPUs", *IEEE Transactions on Parallel and Distributed Systems*, 2012: 23, no. 7, 1255–1266.

[5] G. Soykan, A.J. Flueck and H. Dag, "Parallel-in-space implementation of transient stability analysis on a Linux cluster with infiniband", North American Power Symposium (NAPS), 2012: 1–5.

[6] Z. Huang, R. Diao, S. Jin, Y. Chen, "Predictive dynamic security assessment through advanced computing", IEEE Power and Energy Society General Meeting, 2014, pp. 1–5.

[7] S. Jin, Z. (Henry) Huang, R. Diao, D. Wu and Y. Chen, "Parallel implementation of power system dynamic simulation", IEEE Power and Energy Society General Meeting, 2013, pp. 1–5.

[8] Z. Huang, S. Jin and R. Diao, "Predictive dynamic simulation for large-scale power systems through high-performance computing", in: 2012 SC Companion: High Performance Computing, Networking Storage and Analysis, pp. 347–354.

[9] M. Milton, A. Benigni and J. Bakos, "System-level, FPGA-based, real-time simulation of ship power systems", *IEEE Transactions on Energy Conversion*, 2017: 32, pp. 737–747.

[10] M. Difronzo, M. Milton, M. Davidson and A. Benigni, "Hardware-in-the-loop testing of high switching frequency power electronics converters", IEEE Electric Ship Technologies Symposium (ESTS), 2017, pp. 299–304.

[11] C. Dufour, J. Mahseredjian and J. Belanger, "A combined state-space nodal method for the simulation of power system transient", *IEEE Transactions on Power Delivery*, 2011: 26, pp. 928–935.

[12] Y. Chen, M.G. Fadda and A. Benigni, "Decentralized load estimation for distribution systems using artificial neural networks", *IEEE Transactions on Instrumentation and Measurement*, 2019: 68, pp. 1333–1342.

[13] G. Kron, "Diakoptics; Piecewise Solution of Large-Scale Systems", Schenectady, N.Y., General Electric, 1957–1958.

[14] A. Brameller, M.N. John and M.R. Scott, *"Practical Diakoptics for Electrical Networks"*, Chapman and Hall, London, 1969.

[15] P.B. Johns and R.L. Beurle, "Numerical solution of 2-dimensional scattering problems using a transmission-line matrix", *Proceedings of the Institution of Electrical Engineers,* 1971: 118, no. 9, pp. 1203–1208.

[16] P.B. Johns and M. O'Brien, "Use of the transmission line modeling (t.l.m) method to solve nonlinear networks", *The Radio and Electronic Engineer*, 1980: 50, no. 1/2, pp. 59–70.

[17] C. Christopoulos, "The historical development of TLM [transmission line modelling]", *IEE Colloquium on Transmission Line Matrix Modelling – TLM,* 1991, pp. 1–4.

[18]   S.Y.R. Hui and C. Christopoulos. "A discrete approach to the modeling of power electronic switching networks", *IEEE Transactions on Power Electronics,* 1990: 5, no. 4,pp. 398–403.

[19]   S.Y.R. Hui, K.K. Fung, M.Q. Zhang and C. Christopoulos, "Variable time step technique for transmission line modeling", *IEE Proceedings Science, Measurement and Technology,* 1993: 140, no. 4, pp. 299–302.

[20]   K.K Fung, S.Y.R. Hui and C. Christopoulos, "Concurrent simulation of decoupled power electronic circuits", Fifth European Conference on Power Electronics and Applications, 1993:7, pp. 18–23.

[21]   S.Y.R. Hui, K.K. Fung and C. Christopoulos, "Decoupled simulation of DC-linked power electronic systems using transmission-line links", *IEEE Transactions on Power Electronics,* 1994: 9, no. 1, pp. 85–91.

[22]   S.Y.R. Hui and C. Christopoulos, "Modeling non-linear power electronic circuits with the transmission-line modeling technique", *IEEE Transactions on Power Electronics,* 1995: 10, no. 1, pp. 48–54.

[23]   S.Y.R. Hui and K.K. Fung, "Fast decoupled simulation of large power electronic systems using new two-port companion link models", *IEEE Transactions on Power Electronics,* 1997: 12, no. 3, pp. 462–473.

[24]   A.I. Zecevic and N. Gacic, "A partitioning algorithm for the parallel solution of differential-algebraic equations by waveform relaxation", *IEEE Transactions on Circuits and Systems—I: Fundamental Theory and Applications*, 1999: 46, no. 4, pp. 421–434.

[25]   E. Lelarasmee, A.E. Ruehli and A.L. Sangiovanni-Vincentelli, "The waveform relaxation method for time-domain analysis of large integrated circuits", *IEEE Transactions on Computer-Aided Design,* 1982: 1, pp. 131–145.

[26]   R.A. Saleh and R. Newton, "The exploitation of latency and multirate behavior using nonlinear relaxation for circuit simulation", *IEEE Transactions on Computer-Aided Design of Integrated Circuits and Systems,* 1989: 8, no. 12, pp. 1286–1298.

[27]   P. Cox, R.G. Burch, P. Yang and D.E. Hocevar, "New implicit integration method for efficient latency exploitation in circuit simulation", *IEEE Transactions on Computer-Aided Design of Integrated Circuits and Systems*, 8, no. 10, pp. 1051–1064.

[28]   M. Ilic-Spong, M.L. Crow, M.A. Pai, "Transient stability simulation by waveform relaxation methods", *IEEE Transactions on Power Systems,* 1987: 2, no. 4. pp. 943–949.

[29]   M.L. Crow, M. Ilic, "The parallel implementation of the waveform relaxation method for transient stability simulations", *IEEE Transactions on Power Systems,* 1990: 5, no. 3, pp. 922–932.

[30]   J.E. Schutt-Aine, "Latency Insertion Method (LIM) for the fast transient simulation of large networks", *IEEE Transactions on Circuits and Systems—I: Fundamental Theory and Applications,* 2001: 48, no. 1, pp. 81–89.

[31] S.N. Lalgudi, M. Swaminathan, Y. Kretchmer, "On-chip power-grid simulation using latency insertion method", *IEEE Transactions on Circuits and Systems—I: Regular Papers*, 2008: 55, pp. 914–931.

[32] S.N. Lalgudi, M. Swaminathan, "Analytical stability condition of the latency insertion method for nonuniform GLC circuits", *IEEE Transactions on Circuits and Systems II: Express Briefs*, 2008: 55, pp. 937–941.

[33] Z. Deng, J.E. Schutt-Aine, "Stability analysis of latency insertion method (LIM)", in: IEEE 13th Topical Meeting on Electrical Performance of Electronic Packaging, October 2004, pp. 167–170.

[34] J.E. Schutt-Aine, "Stability analysis of the latency insertion method using a block matrix formulation", Electrical Design of Advanced Packaging and Systems Symposium (EDAPS) 2008, pp. 155–158.

[35] P. Goh, J.E. Schutt-Aine, D. Klokotov, J. Tan, P. Liu, W. Dai, F. Al-Hawari, "Partitioned latency insertion method with a generalized stability criteria", *IEEE Transactions on Components, Packaging and Manufacturing Technology*, 2011: 1, pp. 1447–1455.

[36] K.H. Tan, P. Goh, M.F. Ain, "Voltage-in-current formulation for the latency insertion method for improved stability", IEEE *Electronics Letters*, 2016: 52, pp. 1904–1906.

[37] T. Sekine, H. Asai, "Block latency insertion method (Block-LIM) for fast transient simulation of tightly coupled transmission lines", in: IEEE International Symposium on Electromagnetic Compatibility, 2009.

[38] M. Milton, A. Benigni, "Latency insertion method based real-time simulation of power electronic systems", *IEEE Transaction on Power Electronics*, 2018: 33, pp. 7166–7177.

[39] A. Benigni, A. Monti, R. Dougal, "Latency based approach to the simulation of large power electronics systems", *IEEE Transactions on Power Electronics*, 2014: 29, pp. 3201–3213.

[40] A. Benigni, A. Monti, "A parallel approach to real-time simulation of power electronics systems", *IEEE Transaction on Power Electronics*, 2015: 30, pp. 5192–5206.

[41] M. Marin, A. Benigni, H. Lakhdar, A. Monti, P. J. Collins, "Towards the implementation of a parallel real-time simulator for DSP cluster", in: Proceedings of the International Simulation Multiconference 2012, pp. 1–6.

Chapter 6

Simulation under uncertainty

Matthew Milton[1], Andrea Benigni[2,3] and Antonello Monti[4,5]

6.1 Introduction

Effects of uncertain random elements within systems are expected to be very relevant in future energy systems due to the volatility of pervasive new energy sources (particularly large and small renewable sources, and consumer-owned small sources) and the presence of communication networks (e.g. random communication delays). These uncertain phenomena add to uncertainty sources typical of classical energy networks: load behavior, prices, and components reliability. Uncertainty can be introduced into systems by processes that are too complicated to model deterministically (i.e. exactly), and thus the processes are assumed to be stochastic (i.e. random). Moreover, the uncertainty of a system can come from having limited exact knowledge of elements of said system leading to these elements to be considered random. As uncertain/random quantities can impact the behavior of a system, modeling methods for uncertainty quantification are of great importance and should be considered at all stages of design and operation, including supporting real-time optimization and control.

Simulation under uncertainty can be performed using intrusive and non-intrusive approaches. The main advantage of non-intrusive approaches, such Monte Carlo (MC) for instance, is that no modifications are required on the original deterministic simulation solver of a given system. The main drawbacks of those methods are the very long execution time, and the fully numerical approach that—considering the system as a black box—offers significant less information compared to more flexible analytical approaches. Even if other non-intrusive methods, such as Quasi-Monte Carlo (QMC) and Collocation, have been defined and proved to offer better convergence than MC, they still require multiple sampling of the random space described by the stochastic variables [1].

[1]Department of Electrical Engineering, University of South Carolina, USA
[2]Institute of Energy and Climate Research: Energy Systems Engineering (IEK-10), Juelich Research Center, Germany
[3]Department of Mechanical Engineering, RWTH Aachen University, Germany
[4]Institute for Automation of Complex Power Systems, RWTH Aachen University, Germany
[5]Fraunhofer FIT Center for Digital Energy, Germany

A different approach to simulation under uncertainty is to use an intrusive approach such as the generalized Polynomial Chaos (gPC) [2]. The gPC approach allows obtaining the probability density function (PDF) as well as the statistical moment of system variables by explicit integration of system variables represented as gPC expansion. The use of gPC is typically quite efficient from a computational point of view for smaller systems, and its use in the power system, power electronic area has attracted increasing attention in recent years [3–7]. At the same time, one of the main drawbacks—from an implementation point of view—is the need of reformulating the model equations in the gPC expansion. Non-intrusive approaches based on polynomial chaos have been also proposed in the past as in [8], with limitations similar to QMC and collocation methods.

In this chapter, we present common methods to perform simulation of power (electronic) systems with uncertainty. We begin with an introduction to uncertainty with stochastic quantities and the statistics involved. Then, we present two case studies of how simulation with uncertainty is applied. In following, we present the sample-based Monte Carlo methods for simulation with uncertainty. Finally, we discuss the analytical-based polynomial chaos stochastic approaches.

6.2   Case studies

6.2.1   Case study 1: ship system analysis under uncertainty

In this case study, we show how polynomial chaos can be used to simulate dynamic energy systems. As the first example, we consider a three-phase DC/AC converter. The schematic of the DC/AC converter test case is reported in Figure 6.1; semiconductors are modeled as ideal switching devices and the model parameters are reported in Table 6.1.

We assumed that each of the inverter components (C_{DC1}, C_{DC2}, $L_{filter1}$, $L_{filter2}$, $L_{filter3}$, $R_{L\_filter1}$, $R_{L\_filter2}$, $R_{L\_filter3}$, $C_{filter1}$, $C_{filter2}$, $C_{filter3}$) as well as the loads (R_{load1}, R_{load2}, R_{load3}) are characterized by an independent uncertainty equal to 10% of their nominal value and uniformly distributed around the nominal value. This

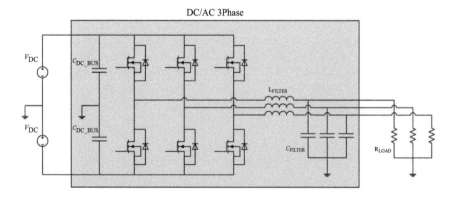

Figure 6.1   DC/AC converter test case

leads to a system characterized by 14 stochastic variables, a second-order approximation has been used.

Figure 6.2(a) reports the expected value for the voltage across the first-phase capacitor filter and the current through the filter inductor of the same phase. Both plots are obtained from the first coefficient of the gPC expansion of the associated variables. In Figure 6.2(b) and (c), histograms representing the pdf of the voltage across the first-phase capacitor filter and the current through the filter inductor at time equal to 0.018 s are reported. Both histograms have been created generating ten thousand samples from the gPC expansion of the associated variables at the selected time.

As the second example, we considered a notional ship power system composed by a generation zone, two propulsion systems and a generic load zone. The single line diagram of the notional ship power system is reported (Figure 6.3), it is

Table 6.1   DC/AC converter parameters

Parameters	Values
V_{DC}	$\pm100\ \Omega$
R_{DC}	$0.01\ \Omega$
L_{filter}	$10\ \text{mH}$
R_{L_filter}	$0.01\ \Omega$
C_{filter}	$20\ \mu\text{F}$
R_{load}	$5\ \Omega$
f_{sw}	$1\ \text{kHz}$

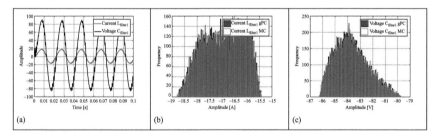

Figure 6.2   (a) DC/AC converter test case average results, (b) histogram of current through $L_{filter1}$ at $T = 0.018$ s, (c) histogram of voltage across $C_{filter1}$ at $T = 0.018s$

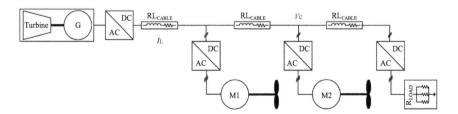

Figure 6.3   Ship power system test case

composed by a generation zone, a 12-kV DC bus, two propulsion systems of 12 MW and 6 MW, respectively, and a 5 MW generic load zone. For the power converters, in this case, an averaged model is proposed. For the two propulsion motors, a fifth-order induction machine model has been used. For the propulsion load, a torque load proportional to the square of the shaft speed is used. For the main generator, a first-order model has been used.

For the ship power system test case, we considered that the parameters of the DC bus cables (R_{cable} L_{cable}), the value of the DC bus capacitors of the zonal load and of the motor drive converters (C_{DC}) and the inertia of both the propulsion induction machines (J_1, J_2) are characterized by an independent uncertainty equal to 10% of their nominal value uniformly distributed around the nominal value. For this second test case, we use gPC to perform a sensitivity analysis to determine the effects on the voltage at the dc bus—VC in figure—and on the total generated current—IL in figure. The sensitivity analysis is performed on a dynamic condition using the method presented [9]. Once the system reaches the steady state, the propulsion load associated with M1 drops of about 1 MW. In Figure 6.4, we report the expected value for the voltage VC, as in the previous case obtained from the first coefficient of the gPC expansion of the associated variables. In the graph, it is clearly visible how the DC bus voltage increases when the propulsion load is reduced and the associated oscillation.

In Figure 6.5, the time-dependent sensitivity indices that indicate how the five considered stochastic variables relate to the DC bus voltage VC are reported. Due

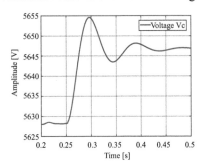

Figure 6.4   Positive DC bus voltage during M1 load change

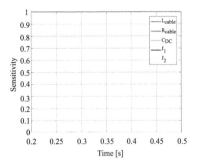

Figure 6.5 V_c sensitivity indices during M1 load change

to the fact that the model is based on averaged converter modeling, in steady state, the voltage is fully determined by the cable resistance (the associated sensitivity index is equal to 1 before the start of the dynamic event). When the propulsion load associated with M1 is reduced, the sensitivity index associated with the resistance has some small fluctuation but it is clearly still fully dominating (see Figure 6.6).

Removing the sensitivity index associated with the resistance, in Figure 6.7, we can see that the sensitivity index associated with the inertia of the induction machine is driving the oscillations, also confirming expectation; as the system tends to a new steady state, the sensitivity index tends toward zero.

In Figure 6.8, we report the expected value for the current I11: as in the previous cases, this is obtained from the first coefficient of the gPC expansion of the associated variables. In the graph, it is clearly visible how the current decreases when the propulsion load is reduced and the corresponding oscillation.

Looking at Figure 6.9—as expected and exactly as in the previous case—before the reduction of propulsive load, the system is in steady and the sensitivity indices associated to the cable inductance L_{cable}, the DC bus capacitors (C_{DC}), and the inertia of both the propulsion induction machines (J_1, J_2) are zero and the index associated with the cable resistance (R_{cable}) is equal to one.

Once the propulsive load reduces, we can see that the inertia of the main induction machine—for which the propulsion load has been reduced—is strongly

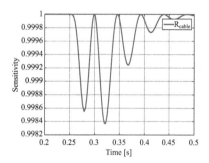

Figure 6.6   Detail of $V_c - R_{cable}$ sensitivity index

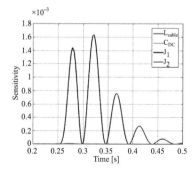

Figure 6.7   All V_c sensitivity indices but $V_c - R_{cable}$

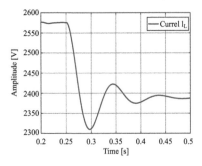

Figure 6.8   Positive DC bus current during M1 load change

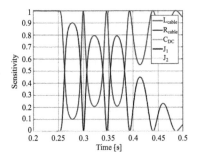

Figure 6.9 I_L sensitivity indices during M1 load change

related together with the cable resistance to the current oscillations. As the system tend to a new steady state, the sensitivity index associated with the inertia tends to zero and the one associated with the cable resistance tends to one again.

6.2.2   Uncertainty sources in the simulation of distribution networks

The growing presence of renewable energy sources (RES) is significantly affecting distribution networks. In particular, this change means also a growing need of active management.

On the other hand, many operators do not have detailed or reliable models of the infrastructure.

From the modeling perspective, distribution networks differ from transmission networks first of all because the high level of imbalance is usually requesting an approach per phase in the simulation scenario. This option makes the model larger and even affected by higher number of parameters.

In the creation of a simulation model, it is then possible to identify a set of uncertainty sources. Starting from the conductors themselves, the first source of uncertainty is the resistance.

Given that in medium voltage and even more in low voltage, the resistance of the cables plays a significant role, the evaluation of the uncertainty in the resistive parameters plays a key role. The same can be said for the inductive component. By looking at real number in realistic applications, the uncertainty on the resistance value is normally bigger than the inductive components.

Furthermore, because the topology may also be not well defined, the parametric uncertainty is not only driven by the value of resistance or inductance per unit of length but even more significantly be the length of the cable itself. In many cases, in effect, the real topology is not available, but it is simply guessed starting from geographical information for some of the key points of the system such as substation locations.

Combining these two factors, it becomes clear that the assessment of the network parameters for a simulation of a distribution network becomes particularly complicated.

The sources of uncertainties, however, are not only limited to the network components. Renewable sources are driven by weather conditions and then an assessment of power injections is particularly complicated and significantly affected by uncertainty.

Loads are also not perfectly predictable and only statistical information are normally available. While on aggregated load, the uncertainty can be rather limited, if we focus the simulation on low voltage scenarios, it becomes extremely difficult to get realistic values for the power consumption of a single household.

On the load side, furthermore, the definition of a model valid for a dynamic simulation is even more complicated. It is not realistic to model each and every device, while the typical approach is to use aggregated models such as the so-called ZIP model. The ZIP model is a combination of an impedance, a current source, and a constant power behavior. Even at this level, the dynamic information is limited and embedded at the end in the impedance value.

Putting all these elements together, it is easy to see that the simulation of a distribution network is normally affected by a number of uncertainties in the order of hundreds.

While this consideration makes the simulation a real challenge, it makes also clear that the deterministic simulation of a limited number of scenarios has very limited value to assess anything regarding the behavior of the grid.

Combining all the effects together, we can easily find in a realistic network that the variation of the voltage profile can be in the range of about 10%. An uncertainty on the voltage profile in this order of magnitude means that assessment of even simple questions such as power quality in the voltage provision become extremely difficult to answer.

Such considerations make clear the need of specific methods for simulation under uncertainty able to extract the key information about the operation of the grid in a reasonable number of scenario simulations.

Situations with more than 100 sources of uncertainties are a typical case showing the so-called course of dimensionality, i.e. the explosion in the number of simulation runs to be executed to get useful results.

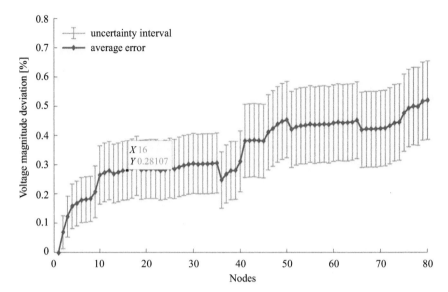

Figure 6.10   Uncertainty on the voltage level for the case of IEEE 123 bus scenario as a consequence of the aging of the cables

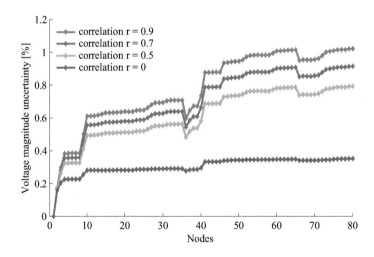

Figure 6.11   Effect of the uncertainty on the cable lengths for different value of correlation for the values in the different phases

On the other hand, as it will be also discussed in the following, methods such as Monte Carlo analysis are always able to address such a large number of uncertainties without an exponential growth in the computational efforts.

Monte Carlo approach is also easy to parallelize for a cluster execution and does not require a special solver but mostly a case generator. These are some of the reasons why, at the end, it is still one of the most applied methods to address challenges such as the one depicted in this introductory example.

Figure 6.10 shows, as an example, how the voltage level is affected by the internal parameters of a line while they are changing because of an aging process.

Figure 6.11, vice versa, analyses the impact of the uncertainty related to the length of the cables and considering different values of correlation for the different phases.

Both calculations have been performed for the reference IEEE 123 buses distribution grids and represent typical effects of uncertainty on the steady-state solution.

6.3 Uncertainty and statistics

When normally performing simulation of a system, we are solving a deterministic model of the system where all quantities (capacitance, resistance, power, etc.) are exactly known, leading to a single exact set of solutions with one outcome. However, with real systems, there is uncertainty in these quantities where we do not know exactly their magnitudes or the complexity to determine the exact magnitudes is impractical. In effect, these uncertain quantities can be considered to be random, or stochastic, where the exact value cannot be precisely determined, but yet can have a regular range of probable values, with each possible value having a probability of occurring. Since these uncertain quantities can have a range of possible values, then in reality, there is a range of possible solution sets for the system in simulation. The idea of simulating with uncertainty is to determine these possible solutions considering the randomness of the quantities.

From probability theory seen in Statistics [10–12], if we were to somehow randomly sample the magnitude of an uncertain quantity for a N number of instances of the quantity, we can collect the samples into a distribution, like as the discrete distribution example in Figure 6.12. In this figure, which is called a histogram, we have a range of discrete magnitudes for the quantity and then for each magnitude, we have a number of occurrences happening for each possible value. The number of occurrences can be analogue to the probability of the magnitude occurring for N samples as greater the occurrence is, the higher chance the associated value will occur. In fact, if we were to divide the number of occurrences of a quantity magnitude by N, then we get a value between 0.0 and 1.0 which indicates the probability of the magnitude for a sample size of N. Note that for a distribution, the magnitudes do not need to be discrete with discrete N samples, but instead can be continuous with infinite N samples within a given range of possible magnitudes (continuous distribution). For a deterministic quantity with no uncertainty, every occurrence of the quantity will be the same magnitude, regardless of the number of samples, so probability will always be 1.0.

From the continuous or discrete distribution of a uncertain quantity with randomness, we can derive what is called a probability density curve which indicates the

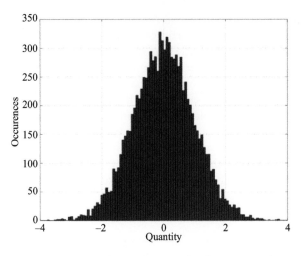

Figure 6.12   Histogram of distribution of a discrete random quantity

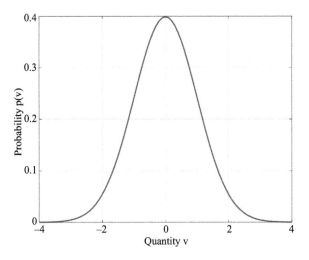

Figure 6.13   PDF curve of a continuous random quantity

probability of a given magnitude of the uncertain quantity; see Figure 6.13, for example. These curves can be described by functions, called PDFs which take as argument the probable magnitude of the quantity v and return the probability of the magnitude, $p(v)$. As such, PDFs describe the distribution of a quantity. Distributions and their PDF curves can take on different shapes depending on the quantity in question, such as the classical continuous bell curve, called normal or Gaussian distribution. Other distributions can include continuous ones such as uniform and Gamma distributions, and discrete distributions such as Poisson and binomial distributions, among many others.

We can calculate what are called statistical moments of a distribution and its PDF for an uncertain quantity, which can tell us information about the distribution of the quantity for analysis, such as shape and spread of the distribution. A common moment is the expected value of the quantity, also known as the mean, which is the average probable magnitude of the quantity to occur; for Gaussian distributions, this is the most probable magnitude. The expected value for respectively a continuous and discrete distribution V are as in (6.1) and (6.2)

$$E[V]_{cont} = \int_S vp(v)dv \tag{6.1}$$

$$E[V]_{disc} = \sum_{i=1}^{N} v_i p(v_i) \tag{6.2}$$

where S is the one-dimensional region from minimum possible value to maximum possible value of v and N is the number of samples in the discrete distribution. To measure the spread of magnitudes in a distribution, we can compute the variance which is the average value of the squared differences of every value in the distribution to from the expected value (6.3) and (6.4)

$$\sigma^2[V]_{cont} = E\left[(V - E[V])^2\right] = \int_S (v - E[V]_{cont})^2 dv \tag{6.3}$$

$$\sigma^2[V]_{disc} = E\left[(V - E[V])^2\right] = \frac{1}{N-1}\sum_{i=1}^{N} (v_i - E[V]_{disc})^2 \tag{6.4}$$

The standard deviation of the distributions is merely the square root of the variance as in (6.5)

$$\sigma[V] = \sqrt{\sigma^2[V]} \tag{6.5}$$

An alternative to variance and standard deviation is the mean absolute deviation (MAD) which is the averaged sum of the absolute differences between each value of the distribution and the expected value, as expressed in (6.6) and (6.7)

$$MAD[V]_{cont} = \int_S |v - E[V]_{cont}|dv \tag{6.6}$$

$$MAD[V]_{disc} = \frac{1}{N}\sum_{i=1}^{N} |v_i - E[V]_{disc}| \tag{6.7}$$

For more information on random variable distributions, PDFs, and moments, it is recommended to review texts on probability theory and statistics, such as [10–12]. These resources and others will go into more detail on the subject and provide overview of other concepts such as law of large numbers and central limit theorem (CLT) which are useful to better understand how various stochastic modeling methods work, such as Monte Carlo.

When performing simulation of a system with uncertainty, we consider that every uncertain (random) quantity of the system has a PDF. The system model is solved considering these PDFs, with the solutions of the system each having their own PDF to define their range of probable outcomes in terms of the quantity PDFs. The representation of these quantity PDFs can be represented in a sampled approach as seen in Monte Carlo or as an analytical Fourier-like approach as seen in polynomial chaos. Each method provides their means to solve system models with uncertainty, as will be described in later sections of this chapter.

6.4   Monte Carlo

We begin our discussion of methods for simulating systems containing uncertainty with the Monte Carlo method. Monte Carlo simulation is a brute force approach to simulate systems with stochastic elements, simply running numerous simulations of the same system for each probable value of the stochastic variables. Under this method, a single model of a system to simulate is defined and the stochastic variables within this model are identified. Then, a finite number of samples are taken from the PDF of the random distribution of each stochastic variable of the system model. For every collection of one sample taken from each and every stochastic variable, a traditional deterministic simulation of the system model is executed, with the stochastic variables set to their sampled value in the collection. After every simulation is performed, the sampled values from each simulation for each solution variable are gathered together to compute the PDF of each solution. These PDFs of the solutions can be used to determine probable values of the solutions as affected by the stochastic variables in the system, as well as statistical moments of these solutions.

Monte Carlo simulation is a useful and robust approach to simulating systems with uncertainty. The approach is simple and does not require intrusively modifying the model of a system to handle the stochastic variables. One merely runs multiple simulations of the same system model with different variable values sampled from the stochastic elements and then just gather the solutions afterward to compute the solution distributions. With parallel computing, such as on graphics processing unit (GPU), field-programmable gate arrays (FPGAs), and computer clusters, numerous simulations can be quickly and readily executed independently from one another to simulate systems with a large number of stochastic elements. Monte Carlo is also quite robust as since the approach does not put requirements on the simulation solver used to simulate the system model, any solver can be used that fits well with the characteristics and dynamics of the system, such as Electromagnetic Transients Program (EMTP), state space, or other types of solvers. Moreover, due to CLT from probability theory, the stochastic solutions of systems simulated with Monte Carlo is expected to converge to correct distributions quickly, regardless of the system model or the distributions of the stochastic variables in said system; so as long as the solver chosen for the system simulation is accurate and the stochastic variables are independent from one another. This characteristic also allows the solution distributions to be analyzed using traditional statistical approaches as often used with normal distributions.

However, the Monte Carlo approach does have its limitations in computation. As multiple simulations need to be performed, computational cost can become expensive. This cost can grow rapidly as the number of stochastic variables grows in amount, as well as how many samples are taken from the distribution of each stochastic variable in the system model. One would need to either have access to multiple parallel computing units to perform all of the simulations quickly, or allow for a large amount of computation time if the simulations are to be executed sequentially on traditional computer hardware (CPUs). Computational cost can be reduced by restricting the number of simulations performed for every collection of stochastic variable samples, or by reducing number of stochastic variables and/or their samples, but accuracy of the solutions can be impacted negatively if the restriction is too aggressive. Approaches to improve computational cost of Monte Carlo while still maintaining (or improving) accuracy and convergence of the method have been proposed via influencing the stochastic inputs to a system model, such as in QMC and Latin hypercube sampling.

6.4.1 Theory

The Monte Carlo approach is a statistical sampling method often used to approximate the solution of a problem by using samples taken from random distributions as inputs to the problem. A common problem is the integration of a given function over a given region [1], such as defined in (6.8):

$$Q = \int_S f(v)dv \tag{6.8}$$

where S is a region of dimension t, f is an integrable function, v is the t-dimensional inputs to f within region S, and Q is the integration solution of f over S. To solve (6.8) for Q using Monte Carlo, we can numerically integrate discretely the function f where the inputs v are randomized samples taken from a probability distribution, as in (6.9):

$$Q_n = \frac{1}{n}\sum_{i=1}^{n} f(v), \quad v = \xi_i \tag{6.9}$$

where n is the number of samples, ξ_i is the ith sample taken from a distribution of independent and identically distributed random values whose dimension is the same as region S and bounded by S, and Q_n is the approximate integration solution for n samples. As n approaches infinity, Q_n computed with Monte Carlo will approach the exact solution of Q (6.10) as defined from law of large numbers from probability theory [13]:

$$\lim_{n \to \infty} Q_n = Q \tag{6.10}$$

Since every sample ξ_i, collected as a n-sized set of samples $\{\xi_i\}_n$ (6.11)

$$\{\xi_i\}_n = \{\xi_1, \xi_2, \ldots, \xi_n\} \tag{6.11}$$

is from a random distribution and due to relationship of (6.10), Q_n is considered the expected (or averaged) value of $f(v)$ under region S, computed from $\{\xi_i\}_n$ (6.12):

$$E[f(v)] = Q \approx Q_n \tag{6.12}$$

where $E[\,]$ is expected (mean) value of its argument. For practical computation, n is set to be finite and as such Q_n computed with Monte Carlo will have error in comparison to Q. The standard deviation error σ_{err} of Q_n from Monte Carlo to the actual solution Q is computed as such (6.13):

$$\sigma_{err} = \sqrt{E\left[(Q - Q_n)^2\right]} \tag{6.13}$$

From CLT in probability theory [13], if the function f has a finite variance as defined by (6.14)

$$\sigma^2(f) = \int_S (f(v) - Q)^2 dx \tag{6.14}$$

then σ_{err} will tend toward

$$\sigma_{err} = \frac{\sigma(f)}{\sqrt{n}} \tag{6.15}$$

as n approaches infinity. As such, the error decreases as the number of samples n is increased, allowing Monte Carlo solution Q_n to converge to exact solution Q.

Example 6.1 Let us solve the integration of the function $e^{(v)}$ over the one-dimensional region $S = [0, 1]$ using the Monte Carlo method:

$$Q = \int_0^1 e^{(v)} dv \tag{6.16}$$

Using traditional analytical approach to solve for Q without Monte Carlo, we find the exact solution to be:

$$Q = e^{(v)}\big|_0^1 = e^1 - e^0 = e - 1 \cong 1.718282 \tag{6.17}$$

To integrate the function with Monte Carlo, we setup the discrete integration like so:

$$Q_n = \frac{1}{n}\sum_{i=1}^{n} e^{v_i} \tag{6.18}$$

If we use $n = 5$ random samples taken from a uniform distribution (where every probable value of the distribution has equal probability) whose range is of [0,1], we can get a sample set such as:

$$\{v_i\}_5 = \{0.7997, 0.5290, 0.8478, 0.4246, 0.1236\} \tag{6.19}$$

From this set, we can solve for Q_5:

$$Q_5 = \frac{1}{5}\left(e^{0.7997} + e^{0.5290} + e^{0.8478} + e^{0.4246} + e^{0.1236}\right) \cong 1.7834 \tag{6.20}$$

Which is close to Q. For other sets of $\{v_i\}_5$ taken from same distribution randomly, Q_5 can also equal 1.8508, 1.6447, 1.3337, 1.7141, etc., showing that Q_5 can deviate from Q randomly from using small n. As n increases, the probable values of Q_n will grow closer to Q consistently for any given $\{v_i\}_n$, reducing error of the method to exact solution.

Try this: Using MATLAB® or Octave, solve for Q_n of this example using $n=10$, 100, 1,000, and 10,000. Note how Q_n converges to Q more consistently as n increases. The rand() function can be used to create the pseudo-random samples $\{v_i\}_n$ from an uniform distribution with values [0,1].

The Monte Carlo approach can be used to solve systems with uncertainty for simulation, where variables of the system have variance described by probability distributions. We can define $f(v)$ as the mathematical model of the system to solve and select v to be a collection of j number of variables and inputs v_i of the system model (6.21):

$$v = [v_1, v_2, \ldots, v_j] \tag{6.21}$$

Then, we can select variables of v to be those with uncertainty with finite variance (stochastic variables) which can be described by a random distribution defined by a PDF. A random sample is taken for each variable in v from their probability distributions to create collection ξ_i and then n number of ξ_i are taken to create a set of samples $\{\xi_i\}_n$. The system model $f(v)$ can be solved for each $v = \xi_i$ to get a collection of solutions y_i, which from solving for every y_i, we get solution set $\{y_i\}_n$ (6.22) and (6.23):

$$y_i = f(\xi_i) \tag{6.22}$$

$$\{y_i\}_n = f\{\xi_i\}_n = \{f(\xi_1), f(\xi_2), \ldots, f(\xi_n)\} \tag{6.23}$$

The set $\{y_i\}_n$ is effectively the sampled random distribution of the solution of f. By applying (6.12) to solutions $\{y_i\}_n$ of f, the approximate expected values of solutions to $f(v)$ can be solved. Other computations can be performed on $\{y_i\}_n$ to determine moments, such as deviation and variance. For stepped time simulations, f can be solved with $\{\xi_i\}_n$ every time step, which $\{\xi_i\}_n$ can be a function of time, to get $\{y_i\}_n$ for each time step.

Example 6.2 Let us say that we want to simulate with the Monte Carlo method a series combination of an inductor L and resistor R with a voltage V across the combination, as part of a larger system model, with L, R, and V being stochastic variables. The differential equation used to model the inductor by its current is:

$$\frac{dI}{dt}(t) = \frac{1}{L}V(t) - \frac{R}{L}I(t) \tag{6.24}$$

We discretize this equation for given dt into a difference equation which can be solved every time step during simulation, using Euler forward discretization here for sheer simplicity:

$$f(v) = y = I(t+dt) = \left(1 - \frac{dtR}{L}\right)I(t) + \frac{dt}{L}V(t) \tag{6.25}$$

$$v = [L, R, V(t), I(t), t] = [v_1, v_2, v_3, v_4, v_5]$$

To solve this model with L, R, and V being stochastic variables with Monte Carlo method, we define the sample set $\{\xi_i\}_n$ to be input to $f(v)$ to get $f\{\xi_i\}_n$, with each ξ_i defined as also a set:

$$\xi_i = v = \{L_i, R_i, V_i(t), I_i(t), t\} \tag{6.26}$$

where t is the same value for every ith sample in a single time step as it is not stochastic, and L_i, R_i, and $V_i(t)$ are ith random samples taken from their respective distributions; $I_i(t)$ is the ith sample of solution distribution of f from previous time step which should be stochastic as well. From this setup, $f\{\xi_i\}_n$ can be solved every time step to get $\{y_i\}_n$ which is the sampled random distribution of $I(t+dt)$. $\{y_i\}_n$ is sampled from as $I_i(t)$ for next time step.

For demonstrating the simulation of this circuit with Monte Carlo method, let us say that $L = 10$ mH, $R = 0.1$, and $V = 10$ V as expected values. Each of these inputs is described in this particular case as having an uniform distribution based on a tolerance of 10% (0.1). If we were to randomly sample the probability distribution of each of these inputs for $n = 10$, we can get the following samples:

$\{\xi_i\}$	ξ_1	ξ_2	ξ_3	ξ_4	ξ_5	ξ_6	ξ_7	ξ_8	ξ_9	ξ_{10}
L_i	0.010024	0.010191	0.0098348	0.010991	0.011	0.010925	0.0095484	0.010988	0.010499	0.0097408
R_i	0.099706	0.09616	0.090995	0.098155	0.11	0.10161	0.1041	0.092494	0.093278	0.094215
V_i	10.345	9.3119	10.9	9.7093	9.4918	11.0	10.535	9.1925	9.2528	10.981

For each sample set ξ_i, we can for a single time step solve for $f(v = \xi_i)$ to get the solution for $y_i = I_i(t+dt)$, and then collect all y_i together into $\{y_i\}_n$ which acts as the distribution for the solution of a single time step. This solution distribution can then be sampled from as input $I_i(t)$ in ξ_i for next time step. Solving $f(v)$ for the time steps at $T=1$ ms and $T=2$ ms for each $v = \xi_i$, we get the following solution distributions, where expected value for $I(1\text{ms})$ is 1 A and $I(2\text{ms})$ is ~2 A.

$\{Y_i\}_n$	y_1	y_2	y_3	y_4	y_5	y_6	y_7	y_8	y_9	y_{10}
$I_i(1\text{ ms})$	1.0321	0.91375	1.1083	0.88337	0.86289	1.0068	1.1033	0.83658	0.88132	1.1274
$I_i(2\text{ ms})$	2.0539	1.8189	2.2063	1.7588	1.7171	2.0043	2.1947	1.6661	1.7548	2.2438

If we ran this simulation where the number of stochastic samples n is large, we can get the following histograms for $I(1\text{ ms})$ and $I(2\text{ ms})$, which have a Gaussian-like distribution as expected from Monte Carlo method:

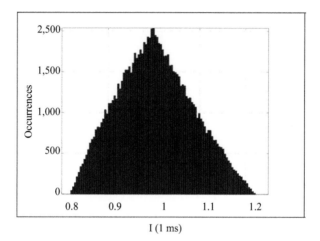

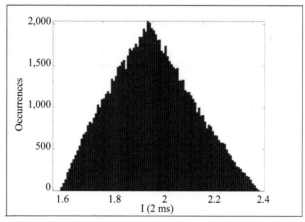

6.4.2 Computation of Monte Carlo simulations

For performing Monte Carlo simulations on system models with uncertainty, the computations of the system model can be done in either of at least two ways which will produce equivalent results: per-sample and vectorized. With per-sample computation, the system model $f(v)$ is solved once for each ξ_i, with each ith instance of the solving of $f(v)$ done separately from the others, like so (6.27):

$$y_1 = f(\xi_1) \tag{6.27}$$

$$y_2 = f(\xi_2)$$

$$\dots$$

$$y_n = f(\xi_n)$$

This computation approach is convenient as it allows the instances to be computed either sequentially in a looped routine or all in parallel across multiple execution units or processes, without modifying the expression of $f(v)$. The approach is also compatible with most computing platforms due to its simplicity.

Another computation approach is vectorized computation. In this approach, the expression (6.23) is computed as one instance, where $\{y_i\}_n$ and $\{\xi_i\}_n$ are expressed as vectors of samples and element-wise matrix calculations are performed to solve $f\{\xi_i\}_n$, like so (6.28):

$$\begin{bmatrix} y_1 \\ y_2 \\ \dots \\ y_n \end{bmatrix} = f\left(\begin{bmatrix} \xi_1 \\ \xi_2 \\ \dots \\ \xi_n \end{bmatrix}\right) \tag{6.28}$$

Some linear algebra software tools such as MATLAB and Octave, and software libraries such as LAPACK and Eigen, are optimized to accelerate such computations, as well as computing hardware including modern GPUs and CPUs. However, this approach much be supported by the computing environment and does not support general parallelization that the per-sample approach allows. Despite which computation approach is used solve Monte Carlo models, the end solutions should be identical. As such, the choice of either approach depends on the computing environment used.

To generate random samples from distributions for Monte Carlo model inputs in practical computing platforms, pseudo-random number generators are used. In common computing environments, these generators are functions, methods, or routines which take a seed value and from that produce a pseudo-random sequence of numbers, typically with uniform probability distribution. Though these generated numbers are actually deterministic, they have enough complexity in their generation to appear random. This pseudo-randomness is often suitable enough for Monte Carlo simulation, especially as true random number generators are typically limited in availability and require separate hardware to create the random values. Many software environments used for simulation and modeling, such as MATLAB, Octave, C++, Python, and Fortran, typically include libraries that provide not only pseudo-random number generators but tools to also produce various common distributions of numbers from these generators (Gaussian, Uniform, Gamma, etc.).

6.4.3 QMC

Monte Carlo is a suitable approach to handle uncertainty in system simulations. However, the traditional Monte Carlo method suffers from often requiring numerous samples from a pseudo-random sequence for the solution of a system to converge to expected values consistently where the error is reduced. This convergence condition is expressed in (6.15) where the error goes down with $\sqrt{n}$ increasing, and seen in Example 6.1. Using a large number of samples helps the method to converge but computing the system model for each sample of the large set raises computational time and/or the parallelized computing resources needed to solve said model. Moreover, the samples are computed in pseudorandom fashion, leading to the solution convergence to vary between identical simulation runs unless the number of samples growly greatly (see Example 6.1).

To solve these limitations of traditional Monte Carlo, approaches have been proposed to alter the method to limit the number of sample inputs needed for a stochastic model so as to not only reduce computational cost but to also accelerate the convergence of the approach. One such altered method is QMC. Instead of using a randomly computed sequence (collection) of samples for input, QMC instead uses a deterministically computed set of inputs which still retain the probability distribution of stochastic variable of a given system. These deterministic set of inputs for QMC are often taken from what are called low-discrepancy sequences (LDSs). LDSs can have same PDFs as the pseudorandom-generated sequences, but values within LDSs are more evenly and deterministically spaced in geometric terms within the region S where from the sequence values are taken, leading to better convergence and lower error.

To understand how QMC improves upon Monte Carlo for convergence, we can look to the discrepancy of the sequences used as an input to the model we wish to solve/simulate. Discrepancy is a measure of how well a set of values of a sequence approximates the volume of a subregion J within a region S [1,13,14]. A common definition of discrepancy is star discrepancy, D_n^*, expressed as in (6.29)

$$D_n^* = \frac{sup}{J \subseteq S} \left| \frac{n_j}{n} - Vol(J) \right| \tag{6.29}$$

where J is a t-dimensional subregion contained within t-dimensional region S, n_j is the number of samples in J, n is the number of samples overall in S, and $Vol(J)$ is the volume of J; the operator sup is shorthand for supremum which is the least element in S that is greater than or equal to all elements in J. J is the subregion of which we wish to integrate within for the purposes of solving a problem under Monte Carlo, containing the values of the input sequence used. The upper bound of the error of Monte Carlo is set as a function of the star discrepancy, the bound defined by the Koksma–Hlawka inequality [1,13], expressed as (6.30)

$$\varepsilon(f) = |Q - Q_n| \leq V(f)D_n^* \tag{6.30}$$

where $V(f)$ is a measure of variation of the function f over a space. As $V(f)$ can be considered a constant for given f, the upper bound for error of Monte Carlo can be lowered by decreasing the value of discrepancy of the input sequence used in a Monte Carlo problem, thereby improving convergence. For traditional Monte Carlo with pseudo-random input sequences, the star discrepancy is (6.31)

$$D_n^* = O\left(\sqrt{\frac{\log\log n}{n}}\right) \tag{6.31}$$

However, the star discrepancy of a sequence can be lower, with the lowest being considered to be (6.32)

$$D_n^* = O\left(\frac{(\log n)^t}{n}\right) \tag{6.32}$$

where t is the dimension of the region the sequence values reside. On comparing (6.31) and (6.32), (6.31) can lead to $1/\sqrt{n}$ rate of converge for error in (6.31) while (6.32) can lead to $1/n$ convergence rate which is faster.

To get discrepancies that approach that of (6.32) for improved convergence rates, we can use deterministically computed LDSs as an input to a Monte Carlo problem to get the QMC approach. Several types of LDSs exist [1,13], including Halton, Faure, Niedereiter, and Sobol sequences, which have pro and cons in regard to their dimensionality for integrating over regions of large dimension t, and convergence rate based on their particular star discrepancy. Full details of these particular LSDs are not provided here for purpose of brevity, but some discussion with references can be found in [1,13–15].

6.5   Polynomial chaos

We now proceed with the discussion of gPC expansion, another approach to model and simulate systems with uncertainty. Unlike sample-based approaches like the Monte Carlo methods which represent the stochastic variables of a system as random samples of probability distributions (a.k.a. PDF), gPC expansion instead represents stochastic variables in an analytical, non-sampled fashion. Each stochastic variable of a system with uncertainty under gPC is expressed as summed infinite series of orthogonal polynomials where each polynomial is scaled by a simple coefficient. The polynomials of the series characterize the distribution shape (normal, uniform, etc.) the stochastic variable PDF takes, while the coefficients characterize the magnitudes of the variable distribution. As stochastic variables are expressed as polynomial series under gPC expansion, instead of as scalar values, the mathematical model of a system to be solved during simulation would need to be modified to consider the arithmetic with these series. As such, the use of gPC expansion to model and simulate systems with uncertainty is intrusive to the expression of the system model.

gPC expansion provides conveniences for efficiently simulating systems with uncertainty. As the polynomials of the gPC expansion series of a stochastic variable are orthogonal to one another, one can define a vector space from the series, where each polynomial coefficient is a coordinate within this space characterizing the particular stochastic variable. These coordinates for each stochastic variable can be grouped together into truncated vectors which arithmetic can be performed on to solve the system model using rudimentary linear algebra algorithms developed and optimized over decades. Using vectorized computing units, such as GPUs and FPGAs, simulations of system models with gPC can be accelerated further. Moreover, for a small number of stochastic variables in a system, the number of coefficients per variable can be low, which allows for computational reductions in comparison to sampled approaches such as Monte Carlo, where numerous samples may need to be computed on for each stochastic variable of a system to achieve similar simulation solution accuracy and convergence as with gPC expansion. However, this advantage over sample-based approaches is lost as the number of stochastic variables grows and the number of coefficients kept for each variable increases, due to the exponential-like growth of the complexity (dimensionality) of the arithmetic with gPC expansion.

Other benefits of gPC expansion include cheap computation of statistical moments from stochastic variables, such as variance and deviation, which can be readily derived from the coefficients rather than the represented distribution directly. Also, as gPC is an analytical approach, the representation of the distribution of stochastic variables has theoretical infinite resolution while sampled approaches typically have finite resolution, though representation accuracy of the gPC approach is bounded when using finite (truncated) number of polynomials for practical computation.

6.5.1 Theory

We proceed now with providing a short review on the theory of gPC expansion. For an exhaustive description of the specifics of this method, refer to [2,12]. As the topic of gPC expansion is highly mathematical, it is recommended the reader be somewhat familiar with basic concepts of linear algebra (bases, vector spaces, projections, orthogonal polynomials, etc.), vector calculus, and probability theory to understand the theory clearly; see [10,13,14,16,17] or similar resources for introduction on these topics.

Under the gPC theory, a random process can be expressed analytically as a spectral expansion of random variables that approximates the random process within the vector space defined with the complete basis of orthogonal polynomials that are in terms of random variables with known PDF. Let X be a continuous random process with a finite second order moment. According to [2,12], X can then be represented with the infinite series expansion in (6.33)

$$X(\theta) = \sum_{i=0}^{\infty} a_i \Phi_i(\xi(\theta)),$$

(6.33)

Table 6.2   Correspondence of continuous distributions to orthogonal polynomials

Distribution	Polynomials	Supported ranges
Gaussian	Hermite	$(-\infty, \infty)$
Uniform	Legendre	$[-1, 1]$
Beta	Jacobi	$[-1, 1]$
Exponential	Laguerre	$[0, \infty)$
Gamma	Generalized Laguerre	$[0, \infty)$

Where θ is the collection of inputs of process X, $\{\Phi_i\}_\infty$ is a set of orthogonal polynomials defined from the Askey hierarchy scheme, a_i is the projection of X on Φ_i, and $\xi(\theta)$ is a collection of artificial random variables serving as arguments to $\{\Phi_i\}_\infty$ in terms of the inputs to X, θ. $\xi(\theta)$ traditionally has PDFs that correspond optimally to the orthogonal polynomials chosen, according to Table 6.2. In simpler terms, the choice of $\{\Phi_i\}_\infty$ defines the probability distribution shape of X and $\{a_i\}_\infty$ defines the probability magnitudes of the distribution.

In the classical use of gPC expansion, the orthogonal polynomials are chosen to match the distribution of X, because the convergence rate of the expansion to expected values is exponentially accelerated in this case. However, according to [12], other polynomials can also provide a valid representation of X, with a slightly impairment to convergence rate. Therefore, it is legitimate to represent any continuous random process with, for example, the Legendre polynomials for simple and practical application of the approach.

For any practical application, the expansion of (6.33) has to be truncated to a finite series. The truncated spectral expansion of a second-order process using gPC is expressed as in (6.34):

$$X_P(\theta) = \sum_{i=0}^{P-1} a_i \Phi_i(\xi(\theta)) \tag{6.34}$$

where P is the total finite number of expansion terms in the series and X_P is the approximation of X. The term P is calculated as such (6.35):

$$P = \frac{(n+g)!}{n!g!} \tag{6.35}$$

where n is the number of stochastic variables in ξ and g is the degree (order) of the orthogonal polynomials base $\{\Phi_i\}_P$. As (6.34) is truncated, the representation of X_p by the gPC expansion will be approximate in comparison to the exact solution of X, with accuracy and convergence to X being a function of P. The standard deviation error between X and X_p can be computed as in (6.36)

$$\sigma_{err} = \sqrt{E\left[(X - X_p)^2\right]} \tag{6.36}$$

with the error converging to zero as P approaches infinity.

6.5.2   *Statistical moments*

We can conveniently compute statistical moments, such as expected value and variance, directly from the coefficients $\{a_i\}_P$ of the gPC expansion of a stochastic process. The expected value of the process is merely the first coefficient of the expansion (6.37)

$$E[X] \approx E[X_P] = a_0 \tag{6.37}$$

The variance of the process is computed as the sum of squares of each coefficient without the first one (6.38):

$$\sigma^2(X) \approx \sigma^2(X_P) = \sum_{i=1}^{P-1} (a_i)^2 \tag{6.38}$$

Standard deviation of the process is merely the square root of (6.38) and (6.39):

$$\sigma(X) \approx \sigma(X_P) = \sqrt{\sum_{i=1}^{P-1} (a_i)^2} \tag{6.39}$$

Mean absolute deviation (MAD) is computed as the sum of the absolute values of the coefficients without the first one (6.40)

$$MAD(X) \approx MAD(X_P) = \sum_{i=1}^{P-1} |a_i| \tag{6.40}$$

6.5.3   *Inner product calculation*

An important operation for performing arithmetic between two stochastic processes in gPC representation and the inner product calculation (generalization of the dot product) between base polynomials under gPC is presented here.

Since all bases of a stochastic process in gPC expansion form are orthogonal polynomials, the inner product between any two polynomials in the expansion can be calculated as the integral of the product of the said polynomials multiplied by a weight function. So considering two orthogonal polynomials Φ_i and Φ_j of a given process, the inner product is obtained using (6.41).

$$\langle \Phi_i, \Phi_j \rangle = \int_{\xi_1}^{\xi_2} \Phi_i(\xi) \Phi_j(\xi) W(\xi) d\xi \tag{6.41}$$

Where the weighting function $W(\xi)$ and the interval region of integration $[\xi_1 \quad \xi_2]$ depend on the specific polynomials considered. In the case of Legendre polynomials, the weighting function W is a constant equal to 0.5 and the interval of integration is equal to the interval $[-1 \quad 1]$.

From a practical point of view, the calculation of the integral in (6.28) can be performed using Gaussian quadrature. Gaussian quadrature rule for the integration of a function in the conventional interval $[-1 \quad 1]$ states that the value of the

integral of interest can be obtained by the summation of a finite number of evaluations of the function itself in selected points. Each evaluation has to be weighted
by a specific coefficient (6.42):

$$\int_{-1}^{1} f(x)dx = \sum_{i=1}^{n} w_i f(x_i) \tag{6.42}$$

The set of points $\{x_i\}_n$ for the evaluation of the function as well as the weights
$\{w\}$ depend on the interval in which the integral needs to be calculated. For the
calculation of an integral in the interval $[-1 \quad 1]$, the Gauss–Legendre quadrature
points need to be used.

6.5.4   Basic algebra using polynomial chaos

To model systems for simulation whose stochastic processes/variables are expressed as gPC polynomial series, we must define basic algebra that allows mathematical operations such as addition and multiplication to be performed between these
process' expansions. A summary of common algebra between two gPC expansion
represented processes of the same polynomial base (vector space) is presented here.
For computational purposes, these operations can be performed directly on the
coefficients of the expansions in a vectorized fashion.

Addition and subtraction

Let us assume that A and B are two stochastic processes of the same gPC
polynomial basis which are to be summed or subtracted to obtain another process C
of the same basis, whose coefficients are labeled, respectively, a, b, and c.
According to the gPC theory, C is computed as (6.43):

$$C = A \pm B = \sum_{i=0}^{P-1} c_i \Phi_i(\xi) = \sum_{i=0}^{P-1} (a_i \pm b_i)\Phi_i(\xi) \tag{6.43}$$

Due to the orthogonality of the basis $\{\Phi_i(\xi)\}_P$, the coefficients of C are calculated as the sum of the coefficients of A and B of the same ith base. Such
operation can be implemented using the pseudo code below:

Create *an object of class pct,*	$c = pct()$
For *each terms of c,*	$k = 1 \ to \ P$
Compute c_k	$value(c_k) = value(a_k) \pm value(b_k)$
End *of iterations*	

Multiplication

To multiply processes A and B to get product C, we perform the operation as in
(6.44) and (6.45):

$$C = AB = \sum_{k=0}^{P-1} c_k \Phi_i(\xi) \tag{6.44}$$

$$c_k = \frac{1}{\langle \Phi_k \Phi_k \rangle} \sum_{i=0}^{P-1} \sum_{j=0}^{P-1} a_i b_j \langle \Phi_k \Phi_i \Phi_j \rangle \tag{6.45}$$

The multiplication can be implemented using the pseudo code below:

Create *an object of class pct,*	$c = pct()$
For *each terms of* c,	$k = 1\ to\ P$
Compute *the inner product*	$D = \langle \phi_k \phi_i \phi_j \rangle$
For *each terms of* a,	$i = 1\ to\ P$
For *each terms of* b,	$j = 1\ to\ P$
Compute *the inner product*	$T = \langle \phi_k \phi_i \phi_j \rangle$
Compute k	$k = \frac{T}{D} * value(a_i) * value(b_i)$
Add k *to* c	$c = c + k$
End *of iterations*	

Division

Let us divide processes A and B to get the quotient C as in (6.46). The division operation can be rearranged into a multiplication as in (6.47) that can be expanded as done before for the multiplication in (6.48):

$$C = \frac{A}{B} \tag{6.46}$$

$$A = CB \tag{6.47}$$

$$a_k = \frac{1}{\langle \Phi_k \Phi_k \rangle} \sum_{i=0}^{P-1} \sum_{j=0}^{P-1} c_i b_j \langle \Phi_k \Phi_i \Phi_j \rangle \tag{6.48}$$

$$\widehat{A} = M_{P,P}\widehat{C} \rightarrow \begin{bmatrix} a_0 \\ \vdots \\ a_{P-1} \end{bmatrix} = \begin{bmatrix} m_{0,0} & \cdots & m_{0,P-1} \\ \vdots & \ddots & \\ m_{P-1,0} & \cdots & m_{P-1,P-1} \end{bmatrix} \begin{bmatrix} c_0 \\ \vdots \\ c_{P-1} \end{bmatrix} \tag{6.49}$$

$$m_{k,i} = \frac{1}{\langle \Phi_k \Phi_k \rangle} \sum_{j=0}^{P-1} b_j \langle \Phi_k \Phi_i \Phi_j \rangle \tag{6.50}$$

$$\widehat{C} = M_{P,P}^{-1}\widehat{A} \tag{6.51}$$

Considering (6.48) for each coefficient k of A, we can rearrange the multiplication in the gPC variables as the multiplication between a square matrix $M_{P,P}$ of dimension $[P, P]$ and a vector composed by the unknown gPC expansion coefficients C to obtain (6.49). The coefficient of matrix $M_{P,P}$ can be calculated using (6.50). To conclude, the unknown gPC expansion coefficients of C can be calculated by (6.51), or more conveniently using one of the many techniques for the solution of a system of linear equations. Considering that $M_{P,P}$ is highly sparse due to the fact that the triple product $\langle \Phi_k \Phi_i \Phi_j \rangle$ is often equal to zero, if P is reasonably high a dedicated algorithm for sparse system may be more convenient. According to what was just explained, the division operation between gPC variables can be implemented using the pseudo code below:

Create an object of class pct, $c = pct()$
For each terms of a, $k = 1$ to P
Compute inner product $D = \langle \phi_k \phi_k \rangle$
For each terms of c, $i = 1$ to P
For each terms of b, $j = 1$ to P
Compute inner product $T = \langle \phi_k \phi_i \phi_j \rangle$
Compute increment $inc = \frac{T}{D} * value(b_i)$
Add increment to $M(k, i)$ $M(k, i) = M(k, i) + add$
End of iterations
Compute c $c = SolveLinearSys(M, a)$

Due to the computational cost of division from having to solve a system of equations for the coefficients of C, it is recommended in general to avoid division with gPC expansion for online simulation, opting instead for multiplication where terms are rearranged offline to avoid the division where possible.

Example 6.3 Let us demonstrate how to perform algebra between two stochastic processes expressed with polynomial chaos expansion, given below. These processes, A and B, are in the same Legendre polynomial basis of order one ($g = 1$), with one stochastic variable ($N = 1$) in a system which to A and B belong; A and B are truncated to two terms ($P = 2$). The process C is the result of the given operation:

$$A = 10\Phi_0(\xi) + 0.5\Phi_1(\xi)$$
$$B = 5\Phi_0(\xi) - 0.1\Phi_1(\xi) \tag{6.52}$$
$$C = c_0\Phi_0(\xi) + c_1\Phi_1(\xi)$$

For each operation, we can create vectors from the coefficients of the processes and perform the algebra on these vectors. Addition and subtraction between A and B is like so:

$$C = A + B \rightarrow \begin{bmatrix} a_0 \\ a_1 \end{bmatrix} + \begin{bmatrix} b_0 \\ b_1 \end{bmatrix} = \begin{bmatrix} 10 \\ 0.5 \end{bmatrix} + \begin{bmatrix} 5 \\ -0.1 \end{bmatrix} = \begin{bmatrix} 15 \\ 0.4 \end{bmatrix} = \begin{bmatrix} c_0 \\ c_1 \end{bmatrix} \rightarrow C$$

$$= 15\Phi_0(\xi) + 0.4\Phi_1(\xi)$$

$$C = A - B \rightarrow \begin{bmatrix} a_0 \\ a_1 \end{bmatrix} - \begin{bmatrix} b_0 \\ b_1 \end{bmatrix} = \begin{bmatrix} 10 \\ 0.5 \end{bmatrix} - \begin{bmatrix} 5 \\ -0.1 \end{bmatrix} = \begin{bmatrix} 5 \\ 0.6 \end{bmatrix} = \begin{bmatrix} c_0 \\ c_1 \end{bmatrix} \rightarrow C$$

$$= 5\Phi_0(\xi) + 0.6\Phi_1(\xi)$$

$$\tag{6.53}$$

To perform multiplication and division between the polynomial processes A and B, we must consider the normalized inner products between the polynomials bases as in (6.37), (6.40), and (6.42), which for the Legendre polynomials with given g, N, and P above, the normalized inner products are given as

$$\frac{\langle \Phi_k \Phi_i \Phi_j \rangle}{\langle \Phi_k \Phi_k \rangle} = \begin{matrix} i = 0 \\ i = 1 \\ j = 0 \quad j = 1 \end{matrix} \begin{bmatrix} p_{(k,0,0)} & p_{(k,0,1)} \\ p_{(k,1,0)} & p_{(k,1,1)} \end{bmatrix} \rightarrow \frac{\langle \Phi_0 \Phi_i \Phi_j \rangle}{\langle \Phi_0 \Phi_0 \rangle}$$

$$= \begin{bmatrix} 1 & 0 \\ 0 & 1/3 \end{bmatrix}, \frac{\langle \Phi_1 \Phi_i \Phi_j \rangle}{\langle \Phi_1 \Phi_1 \rangle} = \begin{bmatrix} 0 & 1 \\ 1 & 0 \end{bmatrix} \tag{6.54}$$

For multiplying A and B to get process C, we do the following as in (6.24)

$$c_0 = a_0 b_0 p_{(0,0,0)} + a_0 b_1 p_{(0,0,1)} + a_1 b_0 p_{(0,1,0)} + a_1 b_1 p_{(0,1,1)}$$

$$c_1 = a_0 b_0 p_{(1,0,0)} + a_0 b_1 p_{(1,0,1)} + a_1 b_0 p_{(1,1,0)} + a_1 b_1 p_{(1,1,1)}$$

$$c_0 = (10)(5)(1) + (10)(-0.1)(0) + (0.5)(5)(0) + (0.5)(-0.1)(1/3)$$

$$= 49.98\overline{3}$$

$$\tag{6.55}$$

$$c_1 = (10)(5)(0) + (10)(-0.1)(1) + (0.5)(5)(1) + (0.5)(-0.1)(0) = 1.5$$

$$C = 49.98\overline{3} \Phi_0(\xi) + 1.5 \Phi_1(\xi)$$

Division is more involved than multiplication, requiring the $M_{P,P}$ matrix to be computed to solve for C. Following (6.29) with $P = 2$, $M_{P,P}$ will be a 2×2 matrix and computed like so, using the normalized inner products

$$M_{2,2} = \begin{bmatrix} m_{0,0} & m_{0,1} \\ m_{1,0} & m_{1,1} \end{bmatrix}$$

$$m_{0,0} = b_0 p_{(0,0,0)} + b_1 p_{(0,0,1)}, \ m_{0,1} = b_0 p_{(0,1,0)} + b_1 p_{(0,1,1)} \tag{6.56}$$

$$m_{1,0} = b_0 p_{(1,0,0)} + b_1 p_{(1,0,1)}, \ m_{1,1} = b_0 p_{(1,1,0)} + b_1 p_{(1,1,1)}$$

$$M_{2,2} = \begin{bmatrix} 5 & -0.0\overline{3} \\ -0.1 & 5 \end{bmatrix}$$

With $M_{2,2}$, we can solve for the coefficients of C like solving a system of equations as in (6.30)

$$\begin{bmatrix} c_0 \\ c_1 \end{bmatrix} = \begin{bmatrix} 5 & -0.0\overline{3} \\ -0.1 & 5 \end{bmatrix}^{-1} \begin{bmatrix} 10 \\ 0.5 \end{bmatrix} = \begin{bmatrix} 2.0009 \\ 0.14 \end{bmatrix} \tag{6.57}$$

$$C = 2.0009 \Phi_0(\xi) + 0.14 \Phi_1(\xi)$$

Computational effort for such operations grows as P increases which makes the algebra tedious to do by hand. As such, linear algebra packages such as MATLAB and Octave can be used to perform the operations in a fast vectorized form.

6.6   Non-intrusive polynomial chaos

6.6.1   Definition of collocation points

In the following, we will present the concept of quadrature-based non-intrusive polynomial chaos. Other non-intrusive concepts, such as sampling and Smolyak sparse grids, which are discussed in more detail in [16], exist as well. However, the most intuitive method for the usage of polynomial chaos with black boxes is based on quadrature rules.

Let $X_1, \ldots, X_n$ be continuous random parameters of a system and Y a target variable which can be derived from the system laws, depending on the random parameters. The system laws can be simple functions, difference equations or differential equations involving the parameters $X_1, \ldots, X_n$. Without loss of generality, we assume that there is only one target variable, since the following procedure can be successively applied to other target variables. Our goal is to compute the PCE of Y, without interfering with the evaluation of the system laws.

As a first step, we will represent all random parameters with the truncated expansions according to paragraph A:

$$X_1(\xi_1) = \sum_{i=0}^{K} a_{1i}\Phi_i(\xi_1), \quad \ldots, X_n(\xi_n) = \sum_{i=0}^{K} a_{ni}\Phi_i(\xi_n). \tag{6.58}$$

Here, Φ_i are the Legendre polynomials and $\xi_1, \ldots, \xi_n$ are uniform random parameters from the $U(-1, 1)$ space. Since each of the expansions only involves one random dimension, the representations can be computed easily. We will concentrate on the case in which $X_1, \ldots, X_n$ are uncorrelated, so $\xi_1, \ldots, \xi_n$ are as well. Additionally, we assume that the random variables are independent. However, it is also possible to transform correlated random variables to uncorrelated variables with adequate transformations, as described e.g. in [16].

Now, let $c_1, \ldots, c_m$ be integration points of a Gaussian quadrature rule on the interval $[-1,1]$, which is also the interval on which the orthogonality of the Legendre polynomials is defined [17]. We compute the following set of points by inserting the integration points into the PCE of the parameters:

$$p_{ij} := X_i(c_j), \quad i = 1, \ldots, n, \ j = 1, \ldots, m. \tag{6.59}$$

These points will be the input for the evaluation of the system laws.

6.6.2   Evaluation

The evaluation step shows the major difference between classical and non-intrusive polynomial chaos. In the classical theory, the full expansions of the parameters are inserted into the system laws which are processed with a Galerkin projection [13], leading to a new, deterministic set of equations which are to be evaluated. In non-intrusive polynomial chaos, however, the evaluation can be conducted in the same way as in Monte-Carlo simulations by inserting samples, which consist of all combinations of the deterministic points p_{ij}. Since the following procedure can be applied to all system equations successively, we will concentrate on a system with one equation. We will denote the solution of the system equation with the input sample $\{p_{1i_1}, \ldots, p_{ni_n}\}$ at the time t as $f(t; p_{1i_1}, \ldots, p_{ni_n})$. In case the system equation consists of a simple function or does not involve time, we can simply assume that t does not occur explicitly in $f(t; p_{1i_1}, \ldots, p_{ni_n})$. We point out once again that this quantity is obtained without interference with the system equation. The only external input to the solution or evaluation process is the sample of points calculated in paragraph B., whereupon the simulation is run independently to give us the output f.

6.6.3 Expansion coefficients of the target variable

The target variable Y depends on n random variables, so its PCE will involve n-dimensional Legendre polynomials $\{\Psi_i\}$, which are tensor products of the one-dimensional polynomials:

$$\Psi_i(\xi_1, \ldots, \xi_n) = \Phi_{i_1}(\xi_1) \otimes \ldots \otimes \Phi_{i_n}(\xi_n). \tag{6.60}$$

According to [14], and remembering that K is the degree of the truncated expansions for the random parameters, the number of summands in the expansion of the target variable is

$$M = \frac{(n+K)!}{n!K!}. \tag{6.61}$$

In Table 6.3, we list the indices $i_1, \ldots, i_n$ of the factors of the first multi-dimensional polynomials in the case that $n = 2$ and $K = 3$, so we have $M = 10$.

Having evaluated the system with the sample sets in paragraph C, we can now compute the PCE coefficient y_i for each Ψ_i with the quadrature rule from which we chose the integration points $c_1, \ldots, c_m$. Since we assume to have independent random variables, we can evaluate each dimension individually. Denoting the

Table 6.3 *Indices of factors for the multi-dimensional polynomials with* $n = 2$ *and* $K = 3$

Index i of multi-dimensional polynomial	i_1	i_2	Comments
0	0	0	$0 + 0 \leq K$
1	0	1	$0 + 1 \leq K$
2	0	2	$0 + 2 \leq K$
3	0	3	.
4	1	0	.
5	1	1	.
6	1	2	.
7	2	0	$1 + 3 > K \Rightarrow$ skip
8	2	1	$2 + 1 \leq K$
9	3	0	$2 + 2 > K$ and $2 + 3 > K \Rightarrow$ skip

corresponding quadrature weights as $w_1, .., w_m$, we find

$$y_i(t) = \frac{1}{\langle \Psi_i^2 \rangle} \cdot \sum_{i_1=0}^{m} \cdots \sum_{i_n=0}^{m} [f(t; p_{1i_1}, \dots, p_{ni_n}) \cdot w_{i_1} \cdots w_{i_n} \cdot \Psi_i(c_{i_1}, \dots, c_{i_n})],$$

(6.62)

so the truncated PCE of Y is now known:

$$Y(t) = \sum_{i=0}^{M-1} y_i(t) \cdot _i(\xi_1, \dots, \xi_n).$$

(6.63)

Now, it is easy to examine the stochastic behavior of the target variable at a desired time instant t_0, which only requires a fast sampling of the $U(-1,1)$ variables $\xi_1, \dots, \xi_n$, inserting t_0 in the coefficients $y_i(t)$.

6.7   Exercises

Monte Carlo exercises
It is recommended to use linear algebra software to solve these exercises, such as MATLAB or Octave.

Exercise MC1

1. Integrate with Monte Carlo approach the following functions over 1D region $S = [0,1]$, for each sample size $n = 10$, 100, 1,000, and 10,000: $\sin v(v)$, $\cos v(v)$, and v^3; use uniform distributed random samples.
2. Repeat problem 1. but over region $S = [-2,3]$
3. Integrate with Monte Carlo approach the following functions over region $S = [0,1]$, for each sample size $n = 10$, 100, 1,000, and 10,000: $\sin^3(v)$, $\cos v(v)\sqrt{1+v^2}$, and $\sqrt{\tan v(v)}$; use uniform distributed random samples.

Exercise MC2

1. Integrate with Monte Carlo approach the following functions $f(u,v)$ over 2D square region $S = [\ 0,1; 0,1\ ]$, for each sample size $n = 10$, 100, 1,000, and 10,000: uv, $\sin(u)\cos v(v)$, and u^{v+1}; use uniform distributed random samples (u,v).
2. Integrate with Monte Carlo approach the following functions $f(u,v,w)$ over 3D cube region $S = [\ 0,1; 0,1; 0,1]$, for each sample size $n = 10$, 100, 1,000, and 10,000: uvw, $\sqrt{u^2+v^2+w^2}$, and $\frac{\sin(u)\cos v(v)}{\tan(w)}$; use uniform distributed random samples (u,v,w).
3. Integrate with Monte Carlo approach the following functions $f(u,v,w,x)$ over 4D hypercube region $S = [\ 0,1; 0,1; 0,1; 0,1]$, for each sample size $n = 10$, 100, 1,000, and 10,000: $(u+v)(w+x)$, and $u\sin(v\cot(w)+x)$; use uniform distributed random samples (u,v,w,x).

Exercise MC3 (a) Simulate the following RLC circuit for three time steps by hand using Monte Carlo, with $dt = 1$ ms and Euler backward discretization, with following parameters; solve for $V_c(t)$ and $I_L(t)$. Use uniform distributions for the uncertain parameters and take $n = 5$ random samples from each parameter distribution. **(b)** Repeat this exercise with software to simulate for 1 s with $n = 10{,}000$ random samples per parameter. Plot the histograms for $V_c(t = 1\mathrm{s})$ and $I_L(t = 1\mathrm{s})$. Plot the expected values of $V_c(t)$ and $I_L(t)$ for $t=[0,1]$ s.

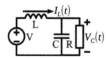

	R	L	C	V	$I_L(t = 0)$	$V_c(t = 0)$
Expected value:	10 Ω	25 mH	16 mF	100 V	0 A	0 V
Tolerance	10%	20%	5%	10%	–	–
Uncertain?	Yes	Yes	Yes	Yes	–	–

Exercise MC4 (Advanced) Simulate the buck converter system below using Monte Carlo, with $dt = 1$ μs, for $T = 3$ s; solve for $V_{ci}(t), I_L(t)$, and $V_{Co}(t)$. Model the converter with state space and switching functions, and use Euler backward discretization. Control the converter by duty (0–1) so output $V_{Co}(t)$ is 480VDC in steady state. Use $n = 10{,}000$ random samples per parameter with uncertainty, taken from uniform distributions. Plot histograms for $V_{ci}(t), I_L(t)$, and $V_{Co}(t)$ at $t = 3$ s. Plot expected values of $V_{ci}(t), I_L(t)$, and $V_{Co}(t)$ for $t = [0,3]$ s.

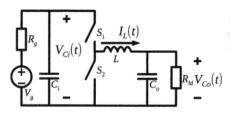

	Rld	L	Ci	Co	Vg	Rg	$I_L(= 0)$	$V_{ci}(t = 0)$	$V_{co}(t = 0)$	$V_{co}(t = \infty)$
Expected value:	5 Ω	25 μH	10 μF	10 μF	1,000 V	0.001	0 A	0 V	0 V	480 V
Tolerance	25%	15%	10%	10%	–	–	–	–	–	–
Uncertain?	Yes	Yes	Yes	Yes	No	No	–	–	–	–

Polynomial chaos exercises
Exercise PCE1

1. Write out the polynomial chaos expansion (PCE) of a generic stochastic process of generic polynomial base where number of stochastic variables is 1 and order/degree is 2.

2. Repeat problem 1 where number of stochastic variables is 2 and order/degree is 3.
3. Repeat problem 1 where number of stochastic variables is 4 and order/degree is 2.

Exercise PCE2 Compute the expected value, variance, standard deviation, and MAD for the following gPC processes.

1. $X = 10\Phi_0 + 0.2\Phi_1$
2. $X = 300\Phi_0 + 25\Phi_1 - 0.2\Phi_2$
3. $X = 256\Phi_0 - 3\Phi_1 + 0.4\Phi_2 - 1.5\Phi_3$

Exercise PCE3

1. Compute the inner product between the Legendre polynomials $P_0(\xi) = 1$ and $P_1(\xi) = \xi$.
2. Compute the inner product between the Legendre polynomials $P_1(\xi) = \xi$ and $P_2(\xi) = \frac{1}{2}(3\xi^2 - 1)$.
3. Compute the triple inner product between the Legendre polynomials $P_0(\xi) = 1$, $P_1(\xi) = \xi$, and $P_2(\xi) = \frac{1}{2}(3\xi^2 - 1)$

Exercise PCE4 Add (A+B), subtract (A−B), multiply (AB), and divide (A/B) the following gPC processes. The base is Legendre polynomials.

1. $A = 1\Phi_0 + 0.4\Phi_1, B = 3\Phi_0 - 1\Phi_1$
2. $A = 62\Phi_0 - 1.0\Phi_1, B = 23\Phi_0 + 0.3\Phi_1$
3. $A = 0.5\Phi_0 + 0\Phi_1, B = 2\Phi_0 + 1.5\Phi_1$

Exercise PCE5 Add (A+B), subtract (A−B), multiply (AB), and divide (A/B) the following gPC processes. The base is Legendre polynomials. (hint: need to compute normalized inner products for $P = 3$)

1. $A = 10\Phi_0 + 2.5\Phi_1 - 0.3\Phi_2, B = 5\Phi_0 - 1.3\Phi_1 - 0.5\Phi_2$
2. $A = 0.33\Phi_0 - 0.66\Phi_1 - 0.16\Phi_2, B = 0.25\Phi_0 + 0.125\Phi_1 + 0.5\Phi_2$
3. $A = 1\Phi_0 + 0\Phi_1 - 2\Phi_2, B = 0\Phi_0 - 5\Phi_1 + 3\Phi_2$

Exercise PCE6 Repeat Exercise MC3 but instead in gPC formulation. Use Legendre polynomials for the uniform distributions and only consider L as uncertain (all other parameters are deterministic with no tolerance). Use a polynomial order/degree of 1 (hint: use linear algebra software to solve the system and remember that gPC coefficients can be grouped into vectors for arithmetic in expressions; see Example 6.3). Plot the expected values of $V_c(t)$ and $I_L(t)$ for $t = [0,1]$ s, but forgo plotting any histograms.

Exercise PCE7 (Advanced) Repeat Exercise MC3 but instead in gPC formulation. Use Legendre polynomials for the uniform distributions and consider all 4 uncertain parameters. Use a polynomial order/degree of 2. Plot the expected values of $V_c(t)$ and $I_L(t)$ for t = [0,1] s. Plot histograms for $V_c(t)$ and $I_L(t)$ at $t = 1$ s (hint: sampled distributions for the histograms can be created from gPC expansions by inserting ξ randomly sampled from normalized uniform distributions into the Legendre polynomials).

Exercise PCE8 (Advanced) Repeat Exercise MC4 but instead in gPC formulation. Use Legendre polynomials for the uniform distributions of the 4 uncertain parameters. Use a polynomial order/degree of 2. Plot expected values of $V_{ci}(t), I_L(t)$, and $V_{Co}(t)$ for $t = [0,3]$ s. Plot histograms for $V_{ci}(t), I_L(t)$, and $V_{Co}(t)$ at $t = 3$ s (hint: sampled distributions for the histograms can be created from gPC expansions by inserting ξ randomly sampled from normalized uniform distributions into the Legendre polynomials).

References

[1] M. Pau, F. Ponci and A. Monti, "Impact of network parameters uncertainties on distribution grid power flow", in: 2019 International Conference on Smart Energy Systems and Technologies (SEST), Porto, Portugal, 2019, pp. 1–6, doi: 10.1109/SEST.2019.8849030.

[2] R. Hogg and A. Craig, *Introduction to Mathematical Statistics*, 4th ed., Macmillan Publishing Co., New York, NY, 1978.

[3] D. Xiu and G. Karniadakis, "The Wiener-Askey polynomials chaos for stochastic differential equations", *SIAM Journal of Scientific Computing*, Vol. 24, pp. 619–644, 2002.

[4] F. Ni, P. Nguyen and J.F.G. Cobben, "Basis-adaptive sparse polynomial chaos expansion for probabilistic power flow", *IEEE Transactions on Power Systems*, Vol. 32, No. 1, pp. 694–704, 2017.

[5] Q. Su and K. Strunz, "Stochastic polynomial-chaos-based average modeling of power electronic systems", *IEEE Transactions on Power Electronics*, Vol. 26, No. 4, pp. 1167–1171, 2011.

[6] H. Li, A. Monti, F. Ponci, W. Li, M. Luo and G. D'Antona, "Voltage sensor validation for decentralized power system monitor using polynomial chaos theory", *IEEE Transactions on Instrumentation and Measurement*, Vol. 60, No. 5, pp. 1633–1643, 2011.

[7] D. Xiu, *Numerical Methods for Stochastic Computations: A Spectral Method Approach*, Princeton University Press, Princeton, NJ, 2010.

[8] P. Prempraneerach, F.S. Hover, M.S. Triantafyllou and G.E. Karniadakis, "Uncertainty quantification in simulations of power systems: multi-element polynomial chaos methods", *Reliability Engineering and System Safety*, Vol. 95, pp. 632–646, 2010.

[9] K. Togawa, A. Benigni and A. Monti, "Advantages and challenges of non-intrusive polynomial chaos theory", in: *ISMC 2011, International Simulation Multiconference*, 27–29 July 2010, The Hague, Netherlands.

[10] R. Larson and B. Edwards, Calculus, 9th ed., *Brooks/Cole Cengage Learning*, Belmont, CA, 2010.

[11] D. Moore and G. McCabe, *Introduction to the Practice of Statistics*, 5th ed., W.H. Freeman and Company, New York, NY, 2006.

[12] T. Wonnacott and R. Wonnacott, *Introductory Statistics*, 2nd ed., John Wiley & Sons Inc., New York, NY, 1972.

[13] H. Niederreiter, "Quasi-Monte Carlo methods and pseudo-random numbers", *Bulletin of the American Mathematical Society*, Vol. 84, No. 6, pp. 957–1041, 1978.

[14] S.P. Oliveira and J.D.S. Azevedo, "A numerical comparison between quasi-Monte Carlo and sparse grid stochastic collocation methods", *CiCp Communications in Computational Physics*, Vol. 12, No. 4, pp. 1051–1069.

[15] A. Singhee and R. Rutenbar, "Why quasi-Monte Carlo is better than Monte Carlo or Latin hypercube sampling for statistical circuit analysis", *IEEE Transactions of Computer-Aided Design of Integrated Circuits and Systems*, Vol. 29, No. 11, 2010.

[16] W. Morokoff and R. Caflisch, "Quasi-random sequences and their discrepancies", *SIAM Journal of Science of Computing*, Vol. 15, No. 6, pp. 1251–1279, 1994.

[17] L. Johnson, R. Riess and J. Arnold, *Introduction to Linear Algebra*, 5th ed., Pearson Education, New York, NY, 2018.

Chapter 7
Simulation language specification—Modelica

Jan Dinkelbach[1], Markus Mirz[1], Antonello Monti[1,2] and Andrea Benigni[3,4]

7.1 Example 1: Simulation of electrical and thermal components considering the impact of a building heating system on the voltage level in a distribution grid

The Modelica language is designed for modeling complex systems with components from multiple domains. We leverage the holistic modeling capability considering electrical and thermal domain. Such considerations are becoming increasingly relevant due to the stronger coupling of both domains in modern distribution grids. In this example, we consider the impact of a building heating system on the voltage level in a low-voltage network.

Figure 7.1 shows the topology of the low voltage grid modeled with a Modelica library for the simulation of power systems called *ModPowerSystems* [1]. The grid consists of 12 buses and has a nominal voltage level of 400 V. A household with a heat pump-driven heating system is connected at node 12, while the other households are considered as classic constant power loads.

Figure 7.2 gives an insight into the model of the heating system installed in the household at node 12. The model is based on components from the libraries *AixLib* [2] and *FastHVAC* [3]. The main components of the heating system are an air-water heat pump, a stratified heat storage, and a radiator. Furthermore, the building is modeled as a single thermal zone demanding thermal energy in order to keep the room temperature at a comfortable level.

Modelica enables to include both electrical and thermal components in one model. A feasible interaction between the components requires an appropriate definition of electrical and thermal connectors that ensure the conservation of power. The electrical connectors include voltage and current as quantities. Instead, thermal connectors involve mass flow rate, temperature, and specific enthalpy as variables.

[1]Institute for Automation of Complex Power Systems, RWTH Aachen University, Germany
[2]Fraunhofer FIT Center for Digital Energy, Germany
[3]Institute of Energy and Climate Research: Energy Systems Engineering (IEK-10), Juelich Research Center, Germany
[4]Department of Mechanical Engineering, RWTH Aachen University, Germany

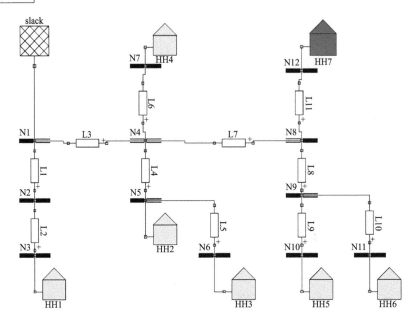

*Figure 7.1   Low-voltage grid including one household with a heating system
driven by a heat pump (marked red)*

We simulate the model of the overall system for an entire year with a time step of 15 min. The resulting voltage at node 12 is depicted as red line in Figure 7.3. It is compared with a simulation result where no heat pump installation was present in the grid. It can be seen that the heat pump installation causes a voltage drop especially during the winter period due to its additional electrical power demand. Instead, the heating system does not impact the voltage significantly during the summer period since the heat-controlled heat pump is mainly switched off.

7.2   Example 2: Static voltage assessment of a distribution grid with high penetration of photovoltaics

Modelica is also suitable for grid simulations and system level considerations. Hence, load flow analyses can be performed using static phasors. For example, these enable the consideration of scenarios with increased penetrations of generation units in distribution grids and the investigation of the impact on the voltage. Here, the IEEE European Low Voltage Test Feeder [4] is implemented in Modelica,

The grid includes 134 buses and 55 households (Figure 7.4). As an example, we consider a scenario in which at 60% of the households, a 5 kW photovoltaic system is installed and we perform a static phasor simulation for a summer day.

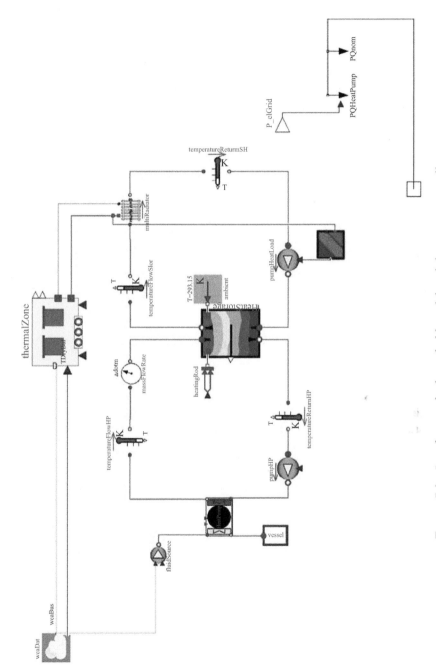

Figure 7.2 Insight into the household model with heat pump installation

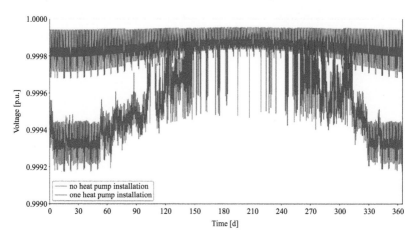

Figure 7.3   Voltage profile at node 12 over one year with one household including a heat pump installation (red) in comparison to a simulation without heat pump installation (blue)

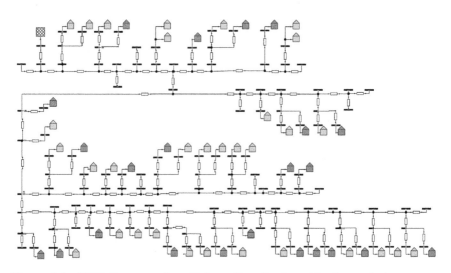

Figure 7.4   IEEE European Low Voltage Test Feeder with photovoltaic installations in 60 % of the households (marked yellow)

Figure 7.5 demonstrates the voltage profile of the node with the highest voltage increase due to distributed generation. In particular, the voltage rises significantly during the midday time because of the high solar radiation. Larger photovoltaic installations or higher photovoltaic penetrations could be easily achieved through model modifications, which would result in a more critical voltage rise.

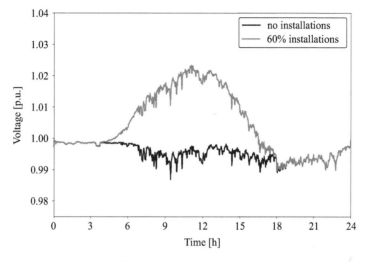

Figure 7.5 Voltage profile at the most critical node during a summer day

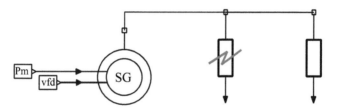

Figure 7.6 Modelica model for simulation of a three-phase fault at synchronous generator terminals

7.3 Example 3: Transient characteristics of synchronous generator models

In this example, we focus on the use of Modelica to analyze the behavior of different models for a specific physical component (Figure 7.6). Modelica enables an easy way to implement models which might have been found in literature. By this, an engineer can investigate the impact of choosing one or the other model by simulating a scenario of interest. At this point, we consider two different models of synchronous generators as described in [5]. Their behavior in the case of a three-phase fault at the generator terminals is simulated in the following.

At first, we consider a detailed model of the synchronous generator assuming one damper winding in the d-axis and two damper windings in the q-axis. The corresponding set of equations as well as the parameters of a 555 MW machine are adopted from [5]. As an extreme case, a three-phase fault of one second is simulated. The diagram of the Modelica model is depicted in Figure 7.6. The synchronous generator model on the left includes the machine equations and receives the mechanical torque P_m as well as the field excitation voltage v_{fd} as input quantities which are applied to the rotor. In a steady state, the generator provides 555kW to a resistive load. The fault at the generator terminals takes place at 0.1s. The simulated currents are shown in Figure 7.7. For the detailed model, the currents during the three-phase fault include an oscillating component and a decaying DC offset.

The voltage equations as introduced in the Modelica code are stated below. We can observe that the code looks very similar to the mathematical descriptions given in literature. The *der* operator is readily applied to obtain the derivative of the magnetic fluxes. The similarity between Modelica code and formulations in literature enables an easy implementation of model descriptions and is one of the major advantages of Modelica.

```
1.  v_d = - R_s*i_d - Psi_q*omega_e +
2.  der(Psi_d)/omega_b;
3.  v_q = - R_s*i_q + Psi_d*omega_e +
4.  der(Psi_q)/omega_b;
5.  v_fd = R_fd*i_fd + der(Psi_fd)/omega_b;
6.  0 = R_kd*i_kd + der(Psi_kd)/omega_b;
```

```
0 = R_kq1*i_kq1 + der(Psi_kq1)/omega_b;
0 = R_kq2*i_kq2 + der(Psi_kq2)/omega_b;
```

We can easily modify the first two voltage equations to obtain another model of the synchronous generator. A common simplified model neglects the derivative terms of the stator flux, which are also known as transformer voltage terms. The

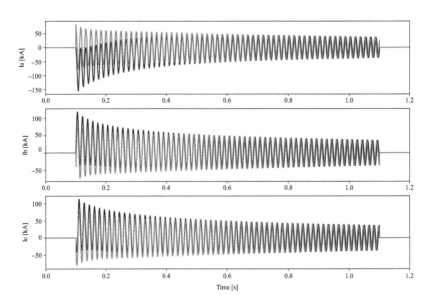

Figure 7.7   Simulated currents occurring during a three-phase fault for two different synchronous machine models (black: detailed model; red: simplified model)

simulation results in Figure 7.7 demonstrate that the simplified model omits the DC offset during the fault occurrence.

7.4   Example 4: Simulation of electrical and mechanical components considering the start of an asynchronous induction machine

Now, we perform a holistic simulation of components from both electrical and mechanical domains. For this, we apply an example model from the Modelica Standard Library, see Figure 7.8, which considers the start of an asynchronous induction machine with squirrel cage. The machine is supplied by a voltage source, while its shaft is connected to a load inertia and a load torque that is quadratically dependent on speed.

The component interaction in each of the domains is achieved through the definition of appropriate connectors. As pointed out in the first example, electrical connectors hold voltage and current as quantities to ensure the conservation of electrical power. Instead, mechanical connectors of rotating components include rotation angle and cut-torque.

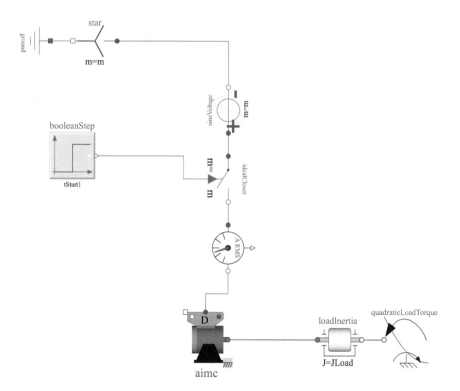

Figure 7.8   Model for simulation of the start of an asynchronous induction machine with squirrel cage (available in the Modelica Standard Library)

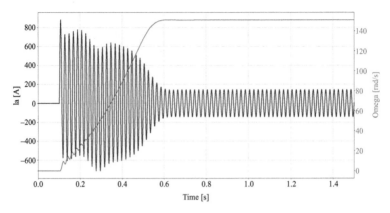

Figure 7.9   Simulation of electrical and mechanical components during start of an asynchronous induction machine (black: supply current phase A, red: angular velocity of load inertia)

The stator side of the asynchronous induction machine is connected through electrical connectors to the supply voltage, while mechanical connectors link machine's shaft, load inertia, and load torque. The resulting load inertia speed-up on the mechanical side and the supply current on the electrical side are depicted in Figure 7.9.

7.5   Introduction to Modelica

Modelica is a modeling language for complex physical systems [6]. Its development has been started in 1996 by an international group aiming at the design of a new modeling language [7]. The main objective was to simplify the exchange of models and to provide a standard format for physical model description [8]. Therefore, the concepts of several modeling languages such as Dymola, ObjectMath, Omola, or VHDL-AMS were taken into account and harmonized.

The first version of the Modelica language was published in 1997. Since 2000, the language has been maintained by the Modelica Association [9], a non-profit organization being responsible for the Modelica Language Specification, which is currently available in version 3.4 [10]. The Modelica Association also manages the development of the Modelica Standard Library which provides various components from several domains.

Modelica's formalism for the specification of the model behavior corresponds with the physical law description typically employed in the literature [6]. Equations determining the physics of a component are introduced in the same manner following the declarative programming paradigm. This simplifies the transfer of knowledge from textbooks into the modeling language and is one of the main characteristics of Modelica.

In particular, it is not required to define for which variables the equations are to be solved. That is, Modelica avoids the a priori definition of input and output variables and does not fix the causality during modeling. When external inputs and outputs are defined, the solution calculation of the equation system is adjusted accordingly. This approach is called acausal modeling and is particularly advantageous to increase the flexibility and reusability of models. Besides, the model description in Modelica is independent from a specific solution methodology. That is, the model does not contain any information on how to solve the equation system describing the physical behavior. In computer science, the paradigm of describing the problem without specifying how to solve it is known as declarative programming. This is in contrast to the imperative programming paradigm, where assignments state how to derive the solution. Through the declarative programing paradigm, Modelica achieves to separate the modeling exercise from the solution exercise.

Modelica is a modeling language for a wide range of physical domains. Hence, it is particularly suitable for modeling heterogeneous physical systems. This has become of special interest in recent times due to the increasing coupling of infrastructure, e.g. in the electrical and thermal domain. For power system simulation, the Modelica Standard Library contains only a limited number of transient and quasistationary models. A more extensive set of models is included in the Modelica PowerSystems library [11]. Also, the *ModPowerSystems* library provides models for different time frames containing static and dynamic phasor as well as electromagnetic transient models [1]. Further libraries for power system modeling are iPSL [12], OpenIPSL [13], and ObjectStab [14].

The Modelica language is supported by several free and commercial programming environments. Typically, these environments include a Modelica compiler or interpreter, an execution, and run-time module, as well as textual and graphical model editors [6]. Freely available and open-source are OpenModelica [15], an environment which is developed by the Open Source Modelica Consortium, and JModelica [16]. A commercial environment with Modelica as its modeling language is Dymola [17]. As mentioned above, the term Dymola was formerly applied for a modeling language from which Modelica has been derived. Further commercial environments are SimulationX [18], MapleSim [19], and Wolfram SystemModeler [20].

7.6   Fundamentals of the Modelica language

We explain the fundamentals of the Modelica language by starting from a simple Hello World example. Then, we describe exemplarily the modeling procedure from the development of a single component model to a composite system model. Besides, we outline how the Modelica code is translated and executed and which further modeling concepts are supported by Modelica. For more details on the Modelica language, the reader might refer to Refs [6,20,21].

7.7   Hello World using Modelica

At first, following the tradition when introducing a computer language, we start with a simple Hello World example using the Modelica language. As the focus of the Modelica language lies on solving equations rather than printing a string, we consider the solution of a simple differential equation:

```
1.  model HelloWorld
2.  Real x(start=1);
3.  equation
4.  der(x) = -x;
5.  end HelloWorld;
```

Modelica provides certain keywords, highlighted in the code in bold, which are reserved. Hence, these keywords may not be used for own naming purposes. The keywords *model* and *end* enclose the description of the model. The *equation* keyword specifies the beginning of the definition of the model behavior. Our example specifies in line 4 a first-order differential equation, which might be described in a book on mathematics by

$$\frac{dx(t)}{dt} = -x(t) \tag{7.1}$$

The differential equation considers x changing its value over time. The first-order time derivative of x is expressed by using the Modelica operator *der()*. The variable x has to be declared before using it in the equation section, which is done in the code in line 2. The declaration involves the specification of the variable's data type. Apart from the data type *Real* used here for floating point values, Modelica provides further basic predefined data types: *Integer, Complex, Boolean, String, Clock,* and *enumeration.*

Furthermore, the solution of the differential equation requires the specification of an initial condition. For the variable x, the initial condition is set during its declaration using a so-called modifier equation. The modifier equation *start = 1,*

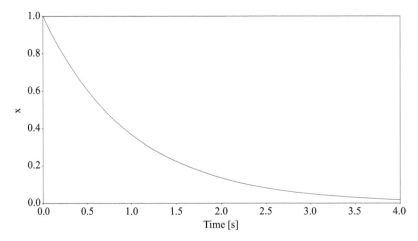

Figure 7.10   Solution of a first-order differential equation obtained with Modelica executing the Hello World example

which is stated in parentheses after the variable name, sets the value of x for the start of the simulation.

The solution of the differential equation can be obtained by executing the model in a suitable Modelica environment.

Figure 7.10 shows the result in a time range from 0 to 4 s

7.8   Electrical component modeling by equations

As shown in the Hello World example, in Modelica, the behavior of a single component can be specified in a declarative manner by the use of equations. Thus, the Modelica language is particularly suitable to implement electrical component models. The following examples are based on the electrical models available from the Modelica Standard Library. First, we consider the model of a resistor:

1. **model** Resistor
2. **extends** Interfaces.OnePort;
3. **parameter** Modelica.SIunits.Resistance R;
4. **equation**
5. v = R*i;
6. **end** Resistor;

In the equation section, we can find the specification of the well-known Ohm's law. The declarations of the variables appearing in the equation, voltage v and current i, are not explicitly part of the *Resistor* model. Instead, the variables are inherited from a base model called *OnePort*. The inheritance from *OnePort* is specified in line 2 using the keyword *extends*. We will have a closer look on the concept of inheritance and the definition of the base model in the next section.

Additionally, in the resistor model, the resistance R is involved its model equation. The same is declared as *parameter* beforehand. A *parameter* in Modelica is a variable that is constant during the simulation, but can be modified, e.g. interactively, immediately before the simulation. The stated type of the parameter is *Modelica.SIunits.Resistance*, which is one of numerous predefined types of the Modelica Standard Library. Additional data types are defined in the Modelica Standard Library for modeling electrical components: *Modelica.SIunits.Voltage, Modelica.SIunits.Current, Modelica.SIunits.ComplexVoltage, Modelica.SIunits. ComplexCurrent, Modelica.SIunits.Angle, Modelica.SIunits.Inductance, Modelica. SIunits. Capacitance, Modelica.SIunits.ActivePower, Modelica.SIunits.ReactivePower,* and *Modelica.SIunits. ApparentPower.*

As further example, we consider the model description of the inductor below. For the inductor, this comprises the implementation of its fundamental differential equation, which involves the time derivative of the inductor current by applying the *der()* operator. The model states the physical behavior of the component without fixing for which variable or with which solution method the equation shall be solved. This points out again that Modelica follows the declarative programing paradigm. Analogously to the resistance, the inductance L in the differential equation is declared before the equation section as *parameter*. Voltage *v* and current *i* are again variables inherited from the base model *OnePort*. In the following section, we have now a closer look on the concept of object-oriented modeling by inheritance and introduce the base model *OnePort*.

```
1. model Inductor
2. extends Interfaces.OnePort;
3. parameter Modelica.SIunits.Inductance L;
4. equation
5. L*der(i) = v;
6. end Inductor;
```

7.9   Object-oriented modeling by inheritance

Modelica is designed as object-oriented programing language to enable the structuring of complex system models and the reuse of model descriptions. Thus, variable declarations, equations and other parts of an existing model (base model) can be inherited to declare a new model (derived model). In Modelica, the derived model specifies its base model by using the keyword *extends*.

Both *Resistor* and *Inductor* model inherit from *OnePort*. According to circuit theory, basic electric elements can be described as one-ports. A one-port consists of two pins and must meet the port condition, which demands that the current flowing out of the port is equal to current flowing in. Basic elements, such as resistor, inductor, and capacitor as well as voltage and current source can be considered as specific kinds of one-ports. Therefore, a one-port is suitable for the definition of a base model. The *OnePort* model as shown below includes a positive and a negative pin. Furthermore, variables *v* and *i* are declared for port voltage and current. The equation section defines how the port quantities are calculated from the pin quantities. The last equation comprises the port condition.

```
 1. partial model OnePort
 2. PositivePin p;
 3. NegativePin n;
 4. Modelica.SIunits.Voltage v;
 5. Modelica.SIunits.Current i;
 6. equation
 7. v = p.v - n.v;
 8. i = p.i;
 9. 0 = p.i + n.i;
10. end OnePort;
```

The *OnePort* model is marked as *partial* in line 1, which denotes that the model requires further specification and is not intended to be instantiated. This corresponds with the concept of abstract classes in other object-oriented languages. In the examples shown above, the models *Resistor* and *Inductor* specify the model *OnePort* by providing additional data, namely resistance and inductance parameters, and characterizing the physical behavior with the corresponding component equation.

7.10   System modeling by composition

To represent an entire system, components of various models can be instantiated and included in one composite system model. Consequently, it emerges the need to specify the interaction between the components. For this, connectors serve as interfaces to the components, thus, allowing them to interact. In the *OnePort* model, *p* and *n* are such interfacing connectors, which are instances from *PositivePin* and *NegativePin*. The declaration of *PositivePin* is the following:

1. **connector** PositivePin
2. Modelica.SIunits.Voltage v;
3. **flow** Modelica.SIunits.Current i;
4. **end** PositivePin;

The *NegativePin* is implemented analogously. In line 1, *PositivePin* is declared as interface model, as the declaration of *PositivePin* uses the keyword *connector* instead of the keyword *model*. If a connection is established between two connectors, their variables will be coupled. *PositivePin* includes a potential variable, the voltage, and a flow variable, the current. The type of coupling depends on whether the corresponding variables are potential or flow variables. Potential variables (without *flow* prefix) lead to an equality coupling. That is, a connection between two or more connectors demands the equality of their potential variables p_i

$$p_1 = p_2 = \ldots = p_n \tag{7.2}$$

Therefore, if we exemplarily connect two instances *pin1* and *pin2* of *PositivePin*, both having voltage *v* as potential variable, this implies

$$pin1.v = pin2.v \tag{7.3}$$

Flow variables are always denoted by the keyword *flow*. For flow variables f_i, a sum-to-zero coupling is defined:

$$f_1 + f_2 + \ldots + f_n = 0 \tag{7.4}$$

In case of the two *PositivePin* instances with the current *i* as flow variable, the sum-to-zero coupling reflects Kirchhoff's current law:

$$pin1.i + pin2.i = 0 \tag{7.5}$$

Analogously, for other domains suitable connectors can be declared. For example, a connector that includes the potential variable *angle* and the flow variable *torque* can be used to model a rotational mechanical movement. Such connector definitions ensure the conservation of power within the specific domain.

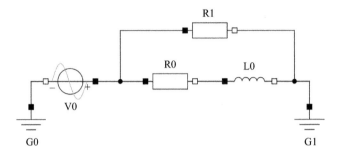

Figure 7.11   Basic electric circuit modeled with components of the Modelica Standard Library

A system such as the basic circuit depicted in Figure 7.11 can be modeled as a composite model. This is due to the fact that Modelica supports hierarchical modeling [8]. The corresponding Modelica code is listed below. The implemented model contains instances of the previously described models *Resistor* and *Inductor* together with further components such as sinusoidal voltage source and ground. The physical connections between the components in the circuit are defined using the *connect* construct in the equation section. Connect-equations are a special kind of equations that establish the equality and sum-to-zero coupling between the corresponding connector variables. Connect-equations are only feasible between instances of connector classes that can be considered as equivalent. During the translation process, they are transformed to normal equations. The overall system of connect and component equations must lead to an equal number of independent equations and variables in order to ensure that the system can be solved.

```
1.  model BasicCircuit
2.  Modelica.Electrical.Analog.Sources.SineVoltage V0 (V=10,
3.  freqHz=50);
4.  Modelica.Electrical.Analog.Basic.Ground G0;
5.  Modelica.Electrical.Analog.Basic.Resistor R0 (R=1);
6.  Modelica.Electrical.Analog.Basic.Inductor    L0 (L=1e-
3);
7.  Modelica.Electrical.Analog.Basic.Ground G1;
8.  Modelica.Electrical.Analog.Basic.Resistor R1 (R=1);
9.  equation
10. connect (V0.n, G0.p);
11. connect (V0.p, R0.p);
12. connect (R0.n, L0.p);
13. connect (L0.n, G1.p);
14. connect (R1.p, R0.p);
15. connect (R1.n, G1.p);
    end BasicCircuit;
```

7.11   Hybrid modeling

The focus of Modelica is on the modeling of continuous physical systems. However, discrete event modeling is supported as well. Like this, hybrid systems including both continuous and discrete dynamics can be modelled. Modelica has four kinds of constructs to represent hybrid models: if-statements to implement discontinuous and conditional models, when-statements that turn active at discontinuities, clocked synchronous constructs and clocked state machines.

The representation of discontinuities is particularly interesting for modern power systems, as they incorporate an increasing amount of power electronic devices and information technology with discrete model parts [22]. In the following, we take a closer look on the use of if-statements for modeling discontinuities. The general form of an if-statement is:

```
if <condition> then
 <statements>
elseif <condition> then
 <statements>
else
 <statements>
end if;
```

As example, we model the discontinuous behavior of a battery with a limited capacity. The definition of such a battery model, here named *SimpleBattery*, is shown below. The battery's capacity is specified by means of the parameter *Cnom* (unit *Wh*). Besides, the battery's state of charge is reflected with respect to the battery capacity by means of the variable *SOC*.

The battery's charging power is controlled according to an input signal, named *PbatIn*. The actual charging power, named *Pbat*, can differ from the control signal due to the capacity limitation. The battery's charging is stopped as soon as the battery leaves the valid operation range, which is defined by a *SOC* between 0 and 1, by enforcing a charging power equal to zero. The corresponding conditional expression in line 7 is based on logical expressions composed by means of the keywords *and* and *or*.

```
1.  model SimpleBattery
2.  parameter Real Cnom;
3.  Real SOC(start=0);
4.  Real Pbat;
5.  Modelica.Blocks.Interfaces.RealInput PbatIn;
6.  equation
7.  if (PbatIn>0 and SOC<1) or (PbatIn<0 and SOC>0) then
8.  Pbat = PbatIn;
9.  else
10. Pbat = 0;
11. end if;
12. der(SOC) = Pbat/(Cnom*3600);
13. end SimpleBattery;
```

The evolution of the *SOC* is described by a first-order differential equation, which implies that the battery's state of charge is the integral over the absorbed power. The SOC's initial condition is set by the modifier equation *start = 0* in the variable's declaration, i.e. we assume that the battery is uncharged at the beginning of the simulation.

The charging control signal *PbatIn* is not given as variable or parameter, but as an interfacing connector, see line 5. Likes this, control signals of any kind can be connected from the outside of the model. The employed connector is of type *RealInput* and defined in the Modelica Standard Library as given below. The connector includes a variable of type *Real* with a fixed causality to *input*, i.e. the information through this variable is unidirectional.

connector RealInput = **input** Real;

As use case, we consider a battery charging according to a specific control input signal. We implement the composite model *BatteryCharging* as listed below. As control input we employ a linearly increasing signal *controlSignal,* see line 3, with a slope of 10 kW/h. Besides, we set the battery capacity to 8 kWh.

```
1.  model BatteryCharging
2.    Battery bat1(Cnom=8000);
3.    Modelica.Blocks.Sources.Ramp
4.      controlSignal(height=100000,duration=36000);
5.  equation
6.    connect(controlSignal.y, bat1.P_batIn);
    end BatteryCharging;
```

As it can be seen in Figure 7.12, at the beginning, the battery charging power follows the input control signal and increases linearly. However, after a charging duration of about 1.26 h, the battery reaches its capacity limit, i.e. the state of charge reaches 100%. Therefore, the battery charging power drops to zero

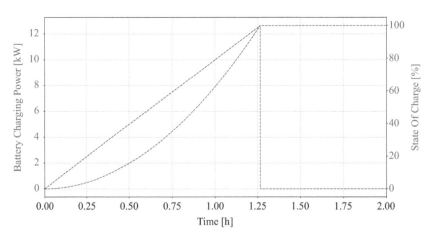

Figure 7.12   Battery charging behavior with linearly increasing charging power and limited battery capacity

according to the if-construct in the *SimpleBattery* model, see line 10. Hence, the *SOC* remains constant in the following.

7.12   Further modeling formalisms

The main modeling formalism in Modelica is the acausal modeling approach. Additionally, Modelica also supports block diagram modeling, which e.g. is employed in Simulink®. For such models, Modelica reserves the class keyword *block*. Block diagram modeling requires already the a priori definition of known and unknown variables and represents the mathematical structure of the problem. Exemplarily, the block diagram of the sample circuit in Figure 7.11 is depicted in Figure 7.13. The block diagram serves to calculate the source current given the source voltage as input. The introduced causality requires distinct models for the resistors, depending on whether the voltage or the current serves as input to the resistor. Hence, we observe that different block representations exist for the two resistors, although their physical behavior is the same. Moreover, changing the topology of the circuit, e.g. introducing a current source instead of a voltage source as input, would require larger modifications of the block diagram. These aspects limit the flexibility and reusability of block diagram models. Besides, the derivation of the block diagram for a given physical system can be rather tedious and non-trivial. Thus, the acausal approach is particularly advantageous for modeling physical systems. However, the block diagram might be useful to demonstrate the mathematical structure of the problem. Furthermore, the block diagram approach is especially suitable for the design of control systems with a fixed signal flow, such as PI controllers.

Except of the acausal and block diagram modeling formalisms, also others such as bond graphs, petri nets and the discrete event system specification (DEVS) might be applied in Modelica for model representation. The DEVS formalism is described in the next chapter.

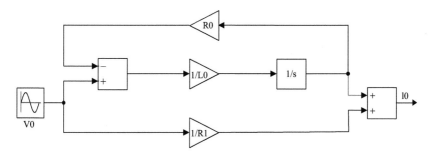

Figure 7.13   Block diagram of the basic electric circuit

7.13   Implementation and execution of Modelica

The typical process of code translation and execution of Modelica code is shown in Figure 7.14. After parsing the Modelica source code, a flat model is obtained e.g. through type checking, the replacement of object-oriented structures and the substitution of connect equations by normal equations. Then, analysis and optimization steps reduce the model to a minimum set of equations. After that, a code generator produces C code which finally is compiled and executed for simulation.

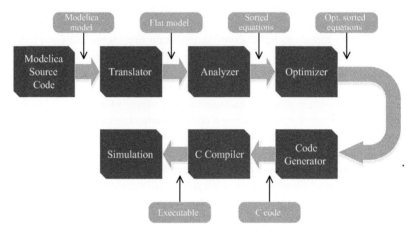

Figure 7.14   Typical process of translation and execution of Modelica code [6]

7.14   Exercises

7.14.1   Task 1

We model and simulate the basic grid shown in Figure 7.15 using the classic phasor representation.

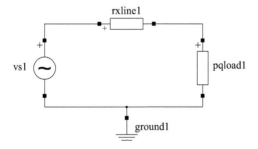

Figure 7.15   Basic grid with ideal voltage source, RX line, and PQ load

Perform the following steps:

1. Define a pin model, namely *ComplexPin,* suitable to perform a phasor simulation.
2. Following the previous step, write a model *ComplexOnePort* serving as base model for power system components that can be considered as one-port.
3. Extending *ComplexOnePort,* develop the following component models:
   (a) An ideal voltage source model, named *VoltageSource,* with *Vmag* and *Vangle* as parameters.

   Hints: Use the constructor *Complex(a,b)* to express a complex number $a + jb$. Besides, you can use the functions *cos()* and *sin()* to define the model equations.
   (b) A line model, named *RXLine,* which reflects resistive and inductive characteristics in the model equation (resistance and inductance in series). As model parameters *length_in_km, r_per_km,* and *x_per_km* shall be implemented.
   (c) A constant power load, named *PQLoad,* with nominal active power *Pnom* and nominal reactive power *Qnom* as parameters. Note that all voltages are phase-to-ground quantities, while the power values are three-phase quantities. In order to ensure that the model converges correctly, insert the following two code lines instead of the usual statement for inheritance from *ComplexOnePort*:

```
extends ComplexOnePort(posPin(v(re(start=Vnom))));
parameter Modelica.SIunits.Voltage Vnom "Nominal voltage";
```
   Hints: You can use the functions *Modelica.ComplexMath.real(), Modelica. ComplexMath.imag(),* and *Modelica.ComplexMath.conj()* to express real and imaginary parts as well as the complex conjugated of a complex quantity.
4. Define an electrical ground model, named *Ground,* which includes one *ComplexPin* instance and defines the potential at this pin to be equal to zero.
5. Using the previously defined component models, write the composite system model *SourceLineLoad* according to Figure 7.15. Use the following parameter values:
   - *Vmag* = 230 and *Vangle* = 0
   - *r_per_km*=0.3, *x_per_km* = 0.1 and *length_in_km* = 1
   - *Vnom* = 230, *Pnom* = 5000 and *Qnom* = 2000
6. Run a simulation of the *SourceLineLoad* model and determine the voltage drop across the line.

7.14.2 Task 2

We extend the battery model *SimpleBattery,* which has been presented above, and analyze its modified charging and discharging behavior.

1. Extend the battery model *SimpleBattery* by adding a limitation of the charging (positive) and discharging (negative) power *Pbat*. For this, introduce a new

parameter *P*max and employ it to limit charging and discharging power by means of an additional if-construct.
2. In the example *BatteryCharging*, set the battery's maximum power *P*max to 6 kW.
3. Run a simulation (with a stop time of 7200 s) and determine
    (a) the time instant as of which the charging power is limited
    (b) the time instant as of which the battery is fully charged

    by plotting the corresponding quantities.
4. Take charging and discharging losses into account by introducing efficiency parameters η_{charge} and $\eta_{discharge}$. Use these parameters in the battery model choosing between suitable differential equations for the *SOC* within an if-construct.
5. Set charging and discharging efficiency to 90% in the *BatteryCharging* example.
6. Re-run the simulation and determine the new time instant as of which the battery is fully charged.

7.15   Exercises—solutions

7.15.1   Task 1—solution

1. *ComplexPin*
    ```
    connector ComplexPin "Pin model for static phasor
    simulation"
        Modelica.SIunits.ComplexVoltage v "Complex poten-
    tial at the pin";
        flow Modelica.SIunits.ComplexCurrent i  "Complex
    current flowing into the pin";
        end ComplexPin;
    ```
2. *ComplexOnePort*
    ```
    partial model ComplexOnePort "Base model for elec-
    trical components that can be considered as one-port"
        ComplexPin posPin "Positive pin of the one-port";
        ComplexPin negPin "Negative pin of the one-port";
        Modelica.SIunits.ComplexVoltage v "Complex voltage
    between posPin and negPin";
        Modelica.SIunits.ComplexCurrent i "Current flowing
    from posPin to negPin";
    equation
        v = posPin.v - negPin.v;
        i = -posPin.i;
        i = negPin.i;
    end ComplexOnePort;
    ```
3. (a) *VoltageSource*
    ```
    model VoltageSource "Ideal voltage source"
        extends ComplexOnePort;
        parameter Modelica.SIunits.Voltage Vmag  "Voltage
    magnitude phase-to-phase RMS amplitude";
    ```

```
    parameter Modelica.SIunits.Angle Vangle "Initial
voltage angle";
    equation
      v = Complex(Vmag*cos(Vangle),Vmag*sin(Vangle));
    end VoltageSource;
```
(b) *RXLine*
```
    model RXLine "Line model composed of resistance and
reactance in series"
      extends ComplexOnePort;
      parameter Real length_in_km "Length of line in km";
      parameter Real r_per_km "Series Resistance per km";
      parameter Real x_per_km "Series Reactance per km";
      Modelica.SIunits.Resistance R "Series Resistance";
      Modelica.SIunits.Reactance X "Series Reactance";
    equation
      R = r_per_km * length_in_km;
      X = x_per_km * length_in_km;
      v = Complex(R,X)*i;
    end RXLine;
```
(c) *PQLoad*
```
    model PQLoad "Constant power load"
      extends ComplexOnePort(posPin(v(re(start=Vnom))));
      parameter Modelica.SIunits.Voltage Vnom "Nominal
voltage";
      parameter Modelica.SIunits.ActivePower Pnom
"Nominal active power";
      parameter Modelica.SIunits.ReactivePower Qnom
"Nominal reactive power";
    equation
      Pnom/3 = Modelica.ComplexMath.real(v*Modelica.
ComplexMath.conj(i));
      Qnom/3 = Modelica.ComplexMath.imag(v*Modelica.
ComplexMath.conj(i));
    end PQLoad;
```
4. *Ground*
```
    model Ground "Electrical ground model"
      ComplexPin pin "Single pin of ground";
    equation
      pin.v = Complex(0,0);
    end Ground;
```
5. *SourceLineLoad*
```
    model SourceLineLoad
      VoltageSource vs1(Vmag=230, Vangle=0);
      RXLine    rxline1(length_in_km=1,    r_per_km=0.3,
x_per_km=0.1);
      PQLoad pqload1(Vnom=230, Pnom=5000, Qnom=2000);
```

```
    Ground ground1;
  equation
    connect(vs1.posPin, rxline1.posPin);
    connect(rxline1.negPin, pqload1.posPin);
    connect(vs1.negPin, ground1.pin);
    connect(pqload1.negPin, ground1.pin);
  end SourceLineLoad;
```

6. Voltage drop across the line: *rxline1.v.re = 2.49083 and rxline1.v.im = −0.144928*

7.15.2   Task 2—solution

1. BatteryPowerLimited

```
  model BatteryPowerLimited
    parameter Real Cnom;
    Real Pbat;
      Real SOC(start=0);
      parameter Real Pmax;
    Modelica.Blocks.Interfaces.RealInput PbatIn;
  equation
      if (PbatIn>0 and SOC<1) or (PbatIn<0 and SOC>0)
then
      if (-Pmax>PbatIn) then
        Pbat = -Pmax;
      elseif (Pmax<PbatIn) then
        Pbat = Pmax;
      else
        Pbat = PbatIn;
      end if;
    else
      Pbat = 0;
    end if;
    der(SOC)= Pbat/(Cnom*3600);
  end BatteryPowerLimited;
```

2. BatteryChargingPowerLimitedExample

```
  model BatteryChargingPowerLimitedExample
    BatteryPowerLimited bat1(Cnom = 8000, Pmax=6000);
    Modelica.Blocks.Sources.Ramp controlSignal(height
= 100000,
      duration=36000);
  equation
    connect(controlSignal.y, bat1.PbatIn);
  end BatteryChargingPowerLimitedExample;
```

3. (a) *ta = 2,160 s, see* Figure 7.16
    (b) *tb = 5,880 s, see* Figure 7.16

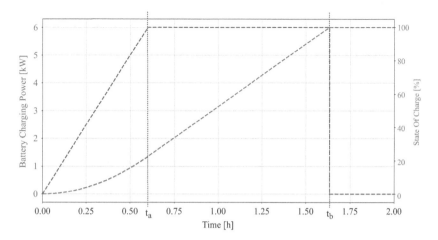

Figure 7.16 Solution Task 2.3

4. BatteryPowerLosses

```
model BatteryPowerLosses
  parameter Real Cnom;
  parameter Real EtaCh;
  parameter Real EtaDisch;
  Real Pbat;
  Real SOC(start=0);
  parameter Real Pmax;
  Modelica.Blocks.Interfaces.RealInput PbatIn;
equation
  if (PbatIn>0 and SOC<1) or (PbatIn<0 and SOC>0) then
    if (-Pmax>PbatIn) then
      Pbat = -Pmax;
    elseif (Pmax<PbatIn) then
      Pbat = Pmax;
    else
      Pbat = PbatIn;
    end if;
  else
    Pbat = 0;
  end if;
  if Pbat>=0 then
    der(SOC)=EtaCh*Pbat/(Cnom*3600);
  else
    der(SOC)=Pbat/(EtaDisch*Cnom*3600);
  end if;
end BatteryPowerLosses;
```

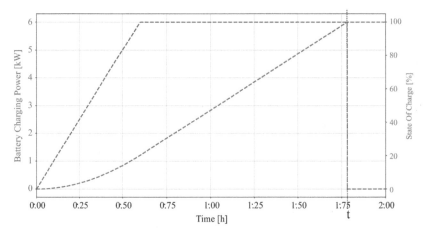

Figure 7.17   Solution Task 2.6

5. BatteryChargingPowerLossesExample

```
model BatteryChargingPowerLossesExample
  BatteryPowerLosses bat1(Cnom = 8000, Pmax=6000,
EtaCh=0.9, EtaDisch=0.9);
  Modelica.Blocks.Sources.Ramp controlSignal(height
= 100000, duration=36000);
equation
  connect(controlSignal.y, bat1.PbatIn);
end BatteryChargingPowerLossesExample;
```

6. $t = 6,413.33$ s, see Figure 7.17

References

[1] Mirz, M., Netze, L., and Monti, A. (2016) "A multi-level approach to power system Modelica models". Presented at the 2016 IEEE 17th Workshop on Control and Modeling for Power Electronics (COMPEL), Trondheim, Norway, 2016.

[2] Müller, D., Lauster, M., Constantin, A., Fuchs, M., and Remmen, P. "An Open-Source Modelica Library within the IEA-EBC Annex 60 Framework". BauSIM 2016, p. 3–9, September 2016.

[3] Stinner, S., Schumacher, M., Finkbeiner, K., Streblow, R., and Müller, D. (2015) "FastHVAC – a library for fast composition and simulation of building energy systems". Presented at the 11th International Modelica Conference, pp. 921–927.

[4] IEEE PES Distribution Systems Analysis Subcommittee Radial Test Feeders. Available: https://ewh.ieee.org/soc/pes/dsacom/testfeeders/.

[5] Kundur, P., Balu, N.J., and Lauby, M.G. (1994) *Power System Stability and Control*, McGraw-Hill, New York; London.

[6] Fritzson, P.A. (2015) *Principles of Object Oriented Modeling and Simulation with Modelica 3.3*, John Wiley & Sons, Hoboken, NJ.

[7] Elmqvist, H. and Mattsson, S.E. (1997) Modelica: An International effort to design the next generation modelling language. In *Proceedings of the 1st World Congress on Systems Simulation WCSS97*, September 1–3, 1997, Singapore.

[8] Elmqvist, H., Mattsson, S.E., and Otter, M. (2001) Object-oriented and hybrid modeling in Modelica. *J. Eur. Systèmes Autom.*, 35 (4), 395–404.

[9] Otter, M. and Elmqvist, H. (2001) Modelica – Language, Libraries, Tools, Workshop and EU-Project RealSim.

[10] Modelica Association (2017) Modelica – A Unified Object-Oriented Language for Systems Modeling – Language Specification – Version 3.4.

[11] Franke, R. and Wiesmann, H. (2014) Flexible modeling of electrical power systems – the Modelica Power Systems library. In *Proceedings of the 10th International Modelica Conference*, pp. 515–522.

[12] Bogodorova, T., Sabate, M., León, G., *et al.* (2013) A Modelica power system library for phasor time-domain simulation. *IEEE PES ISGT Europe 2013*, pp. 1–5.

[13] Vanfretti, L., Rabuzin, T., Baudette, M., and Murad, M. (2016) iTesla Power Systems Library (iPSL): a Modelica library for phasor time-domain simulations. *SoftwareX*, 5, 84–88.

[14] Mats Larsson ObjectStab – a Modelica library for power system stability studies. In *Modelica Workshop 2000 Proceedings*.

[15] Open Source Modelica Consortium (2017) OpenModelica User's Guide.

[16] Åkesson, J., Årzén, K.-E., Gäfvert, M., Bergdahl, T., Tummescheit, H. (2010) Modeling and optimization with Optimica and JMo and delica.org—languages and tools for solving large-scale dynamic optimization problems. *Comput. Chem. Eng.*, 34 (11), 1737–1749.

[17] Wiström, U. (2016) Dymola 2017 – Dynamic Modeling Laboratory (Volume 1).

[18] ESI ITI GmbH SimulationX. Available: https://www.simulationx.com/simulation-software.html.

[19] Maplesoft MapleSim – Advanced System-Level Modeling & Simulation. Available: https://www.maplesoft.com/products/maplesim/.

[20] Wolfram Wolfram SystemModeler: Modellieren, Simulieren & Analysieren. Available: http://www.wolfram.com/system-modeler/.

[21] Tiller, M. (2001) *Introduction to Physical Modeling with Modelica*, Kluwer Academic Publishers, Boston, MA.

[22] Palensky, P., Widl, E., and Elsheikh, A. (2014) Simulating cyber-physical energy systems: challenges, tools and methods. *IEEE Trans. Syst. Man Cybern. Syst.*, 44 (3), 318–326.

Chapter 8
Dynamic phasors

Jan Dinkelbach[1], Markus Mirz[1] and Antonello Monti[1,2]

8.1 Simulation examples

8.1.1 Synchronous generator three-phase fault

The following simulation represents a three-phase fault for the synchronous generator as described in Kundur [1] (Example 3.1). The simulation is conducted with dynamic phasor (DP) variables shifted by the nominal frequency of 60 Hz. Figure 8.1 shows the currents at the synchronous generator terminal. The fault occurs at 0.1 s and is cleared at 0.2 s. In this simulation, the clearing is instantly without waiting for a zero crossing of the current.

From Figure 8.2, it is visible that the rotor frequency of the generator is indeed varying as the faults occurs. Although the shifting frequency of the current phasors

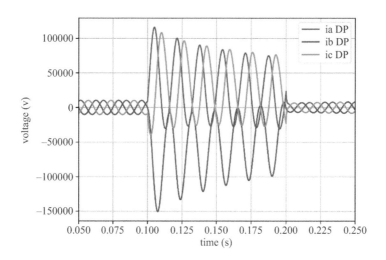

Figure 8.1 Synchronous generator three-phase fault currents

[1]Institute for Automation of Complex Power Systems, RWTH Aachen University, Germany
[2]Fraunhofer FIT Center for Digital Energy, Germany

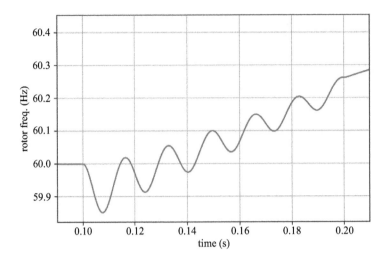

Figure 8.2   Synchronous generator rotor speed

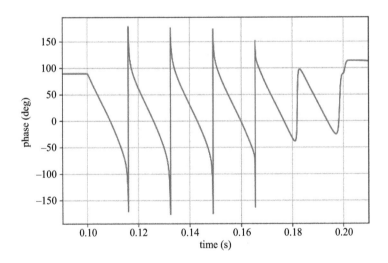

Figure 8.3   Phase variation of dynamic phasor current variable

is 60 Hz, the phasors are not fixed in terms of frequency. The frequency variation is correctly represented, which is visible in the varying phase of the phasor variables depicted in Figure 8.3.

8.1.2   *Grid simulation using diakoptics*

This simulation example demonstrates a combination of dynamic phasors and diakoptics applied to a large network synthesized from several copies of the WSCC

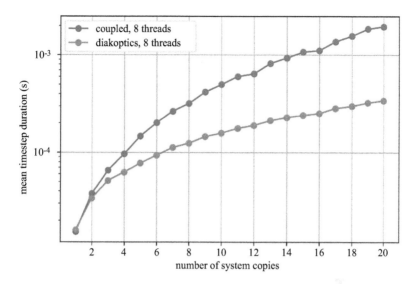

Figure 8.4   Mean computation time per simulation step for varying system size and comparing coupled and diakoptics-based simulation

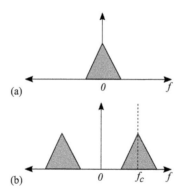

Figure 8.5   Baseband (a) and bandpass (b) signal

9-bus system. The copies are interconnected in a ring type of topology and up to 20 copies are used in this example.

The diakoptics method allows the solver to parallelize the solution of subsections of the entire system model. However, it should be noted that the subsection computation is not completely independent since the diakoptics method results in an additional network consisting of the elements removed between subsections.

Figure 8.4 compares the diakoptics and coupled system computation time per simulation step. The coupled system simulation does not take advantage of splitting at all. Both simulations may use up to 8 threads and parallelize the computation of component states before and after the network solution as defined by the modified nodal analysis scheme including resistive companion models.

Considering the interval between 10 and 20 copies, the computation time for the coupled simulation grows quadratically, whereas the diakoptics simulation grows approximately linearly. The employed solver decomposes the system matrix and subsection matrices before running the actual simulation in order to reduce the computation time.

8.2   Introduction

Dynamic phasors were initially developed for power electronics analysis [2] as a more general approach to averaging than state-space averaging. Later, the concept was extended to power systems analysis [3] to overcome the quasi-stationary assumption of traditional phasors. The authors in [4,5] describe the use of dynamic phasors for power system simulation. In [4], the authors combine the dynamic phasor approach with the Electromagnetic Transients Program (EMTP) simulator concept, which includes modified nodal analysis (MNA).

Still, dynamic phasors are used to construct efficient models for the dynamics of switching gate phenomena with a high level of detail as shown for example in [6,7]. Further research topics include fault and stability analysis under unbalanced conditions as presented in [8,9]. Also rotating machine models have been developed in dynamic phasors. While the authors of [10,11] study Doubly-Fed Induction Generator (DFIG) wind turbines, Refs. [12,13] cover synchronous generator models.

8.3   Comparison to electromechanical simulation

The traditional electromechanical approach for power system stability analysis is based on a 50/60 Hz steady-state phasor representation of the electrical power network. The assumption is that the fundamental frequency of the power system is not deviating much from the nominal frequency even under disturbances. However, this assumption holds true only if the system is mainly powered by large hydro or thermal synchronous generators because they can compensate for sudden changes in electrical power demand due to their large inertia.

The problem of this approach is that it does not account for larger frequency deviations that occur when an increasing amount of energy is produced by renewable energy sources that are interfaced to the grid through power electronics. These renewable energy sources do not inherently feature large inertia and so far cannot substitute large synchronous generators in this regard. Although research is undergoing to provide synthetic inertia by a combination of renewable energy sources and energy storages, renewable energy sources without synthetic inertia provision are still predominant.

In contrast to that, dynamic phasors do not assume a fixed system frequency and, as described in [5], several dynamic phasors with different harmonics can be combined to include not just the fundamental system frequency but also harmonics. The latter can be useful to describe the impact of power electronics on the power system in a more detailed way.

8.4 Bandpass signals and baseband representation

Although dynamic phasors are introduced as power system modeling tool, it should be noted that the concept is also used in other domains, for example, microwave [14,15] and communications engineering [16,17]. In these domains, the approach is often denoted as base band representation or complex envelope [19]. Another related term in power systems is shifted frequency analysis (SFA) [4].

In the following, the general approach of dynamic phasors for power system simulation is explained starting from the idea of bandpass signals since the 50 Hz or 60 Hz fundamental and small deviations from it can be seen as such a bandpass signal. Besides, higher frequencies, for example, generated by power electronics can be modelled in a similar way.

Dynamic phasors are a useful tool, especially, if the signal of interest is a bandpass signal. This means that the signal has a spectrum far from 0 Hz. The center frequency f_c around which the spectrum is located is much higher than the bandwidth of the bandpass signal. The bandwidth of a bandpass signal is the range of positive frequencies, which are present in the signal.

If the bandwidth of a signal is rather small compared to the center frequency, the simulation of such a signal would be very inefficient since the integration time step is determined by the center frequency while the simulation time has to be sufficiently long to notice the modulated signal. Therefore, it would be better to split the bandpass signal into a low-pass component and the center or carrier frequency of the modulated signal. Then, the simulated variable could represent the low-pass signal and in a post processing step of the simulation, the carrier frequency is added. This low-pass signal is denoted baseband signal since its center frequency is 0 Hz.

Figure 8.6 shows two examples of bandpass signals, an amplitude modulated signal and a phase modulated signal. The amplitude modulation affects the envelope function of the signal, which is plotted in red, whereas the phase modulation results in a constant envelope signal.

From Figure 8.7, it can be seen that the spectrum of both signals is centered at 50 Hz and limited in bandwidth. Mathematically, real-valued bandpass signals can be expressed as follows:

$$x_{bp}(t) = A(t)\cos\left(2\pi f_c t + \theta(t)\right) \tag{8.1}$$

Here, $A(t)$ describes the amplitude modulation whereas $\theta(t)$ describes the phase modulation, which can also be used for frequency modulation. f_c is the carrier frequency of the bandpass signal. Equation (1) can be rewritten as follows:

$$x_{bp}(t) = A(t)\cos\left(\theta(t)\right)\cos\left(\omega_c t\right) - A(t)\sin\left(\theta(t)\right)\sin\left(\omega_c t\right)$$

$$= x_{bp,I}(t)\cos\left(\omega_c t\right) - x_{bp,Q}(t)\sin\left(\omega_c t\right) \tag{8.2}$$

where

$$x_{bp,I}(t) = A(t)\cos\left(\theta(t)\right) \tag{8.3}$$
$$x_{bp,Q}(t) = A(t)\sin\left(\theta(t)\right)$$

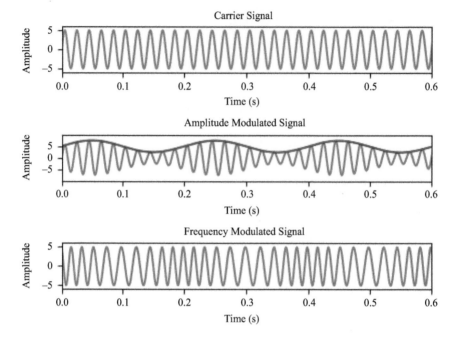

Figure 8.6   Amplitude and frequency modulated carrier signal (bandpass) and modulating signal (baseband)

Since $x_{bpI}(t)$ is in the direction of $\cos(\omega_c t)$ it is called the *in-phase* component, whereas $x_{bpQ}(t)$ is called the *quadrature* component. $x_{bpI}(t)$ and $x_{bpQ}(t)$ contain all information about the bandpass signal $x_{bp}(t)$ except for the carrier frequency f_c. If the carrier frequency is known, the bandpass signal can be reconstructed from the complex low-pass or baseband signal:

$$X_{bb}(t) = x_{bp,I}(t) + jx_{bp,Q}(t) = A(t)e^{j\theta(t)} \tag{8.4}$$

$X_{bb}(t)$ is called the complex envelope of the real signal $x_{bp}(t)$. In power systems and in the following sections, the complex envelope is often denoted as dynamic phasor. In the frequency domain, the dynamic phasor representation is equal to the original signal except for a frequency shift as presented in Figure 8.8. To retrieve the real-valued signal from the dynamic phasor $X_{bb}(t)$, it is sufficient to shift the phasor by the carrier frequency and consider only the real part:

$$
\begin{aligned}
x_{bp}(t) &= \mathrm{Re}\{X_{bp}(t)\} \\
&= \mathrm{Re}\{X_{bb}(t)e^{j\omega_c t}\} \\
&= \mathrm{Re}\{A(t)e^{j(\omega_c t + \theta(t))}\} \\
&= A(t)\cos(\theta(t))\cos(\omega_c t) - A(t)\sin(\theta(t))\sin(\omega_c t) \\
&= A(t)\cos(2\pi f_c t + \theta(t))
\end{aligned}
\tag{8.5}
$$

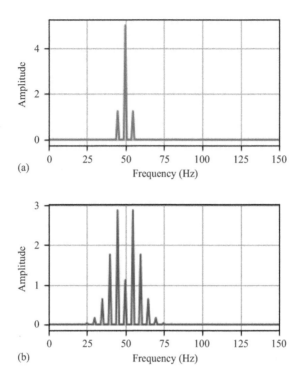

(a)

(b)

Figure 8.7   Positive frequency spectrum of amplitude modulated signal (a) and phase modulated signal (b)

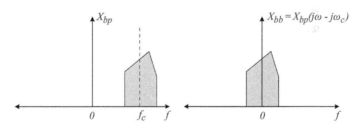

Figure 8.8   Frequency spectrum of a bandpass signal and its baseband representation

In the following, we adopt the nomenclature that is commonly used for dynamic phasors in literature where $\langle x \rangle$ is the dynamic phasor representing a signal x.

8.5   Extracting dynamic phasors from real signals

Different approaches can be applied to retrieve a dynamic phasor representation from a real signal. One option is to calculate the dynamic phasor from the complex analytic signal, which is common in signal processing. The complex analytic signal $x_a(t)$ has the same positive spectrum as $x(t)$ but it is missing the redundant negative frequency part of the real-valued signal.

For the signal considered before $x(t) = x_{bp}(t)$ defined in (1), $x_a(t)$ can be calculated by using the Hilbert transform of $x(t)$ and considering Bedrosian's theorem about the Hilbert transform of a product [18]:

$$
\begin{aligned}
x_a(t) &= x(t) + j\mathcal{H}\{x(t)\} \\
&= x_I(t)\cos(\omega_c t) - x_Q(t)\sin(\omega_c t) + j(x_I(t)\sin(\omega_c t) + x_Q(t)\cos(\omega_c t)) \\
&= (x_I(t) + jx_Q(t)) \cdot (\cos(\omega_c t) + j\sin(\omega_c t)) \\
&= A(t)e^{j\theta(t)}e^{j\omega_c t}
\end{aligned}
\tag{8.6}
$$

Bedrosian's theorem is important at this point since it says that the Hilbert transform of a product of two signals, where one of them is a low-pass signal, equals the product of the low-pass signal and the Hilbert transform of the higher frequency signal:

$$
x_a(t) = A(t)e^{j\theta(t)}e^{j\omega_c t} = \langle x \rangle e^{j\omega_c t}
\tag{8.7}
$$

So $\langle x \rangle(t) = x_a(t)e^{-j\omega_c t}$ is the dynamic phasor, which is shifted in frequency with respect to $x_a(t)$.

Since the shift in frequency only decreases the maximum frequency present in the dynamic phasor if the carrier frequency is larger than the bandwidth of the bandpass signal, dynamic phasors are only efficient in simulation if the bandwidth of the bandpass signal is smaller than its carrier frequency. Then, the baseband signal requires a smaller sampling rate to be represented correctly according to the sampling theorem.

If the bandwidth requirement is fulfilled, we can basically treat an AC signal as a DC signal without losing its dynamic properties as it is the case when using static complex phasors in power system analysis. Instead of fixing the frequency, the signal is shifted by the system frequency, e.g. 50 Hz. The fundamental frequency of power systems is normally varying in a region close to the nominal system frequency. Hence, the bandpass limitation is fulfilled.

This property is very important in the application of real-time simulation since the round-trip time (RTT) between two simulators in different locations can be very significant. The default time step of 50 μs, used by many commercial real-time simulators, does not allow a data exchange between the simulators for every simulation step without compensation for the communication delay.

Time domain variables may include multiple components in the frequency domain as depicted in Figure 8.9. In this case, a single time domain variable can be approximated by several dynamic phasors of different harmonics, each of these shifted by their center frequency:

$$
x(\tau) = \text{Re}\left\{\sum_k \langle x \rangle_k(t)e^{jk\omega_c(\tau)}\right\}
\tag{8.8}
$$

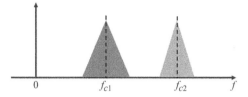

Figure 8.9   Frequency spectrum of a multicomponent signal

This means that the signal can be represented by a Fourier series where the kth Fourier coefficient or dynamic phasor is calculated according to

$$x_k(t) = \frac{1}{T} \int_{t-T}^{T} x(\tau)e^{-jk\omega_c(\tau)}d\tau \qquad (8.9)$$

where $\tau \in (t - T, t]$.

8.6   Modeling dynamic systems using dynamic phasors

Often simulations include dynamic models include differential equations, which need to be expressed in terms of dynamic phasors to take advantage of the approach. Assuming an analytical representation $x_a(t)$ of a real signal $x(t)$,

$$x_a(t) = x(t) + \mathcal{H}\{x(t)\}x_a(t) = A(t)\,e^{j\varphi(t)} \qquad (8.10)$$

this signal can be shifted in the frequency domain by an arbitrary frequency ω_c:

$$
\begin{aligned}
x_a(t) &= A(t)e^{j\varphi(t)-j\omega_c t+j\omega_c t} \\
&= A(t)e^{j(\varphi(t)-\omega_c t)}\,e^{j\omega_c t} \\
&= A(t)\,e^{j\theta(t)}\,e^{j\omega_c t} \\
&= \langle x \rangle(t)\,e^{j\omega_c t}
\end{aligned}
\qquad (8.11)
$$

where

$$\theta(t) = \varphi(t) - \omega_c t \qquad (8.12)$$

$\langle x \rangle(t)$ is the dynamic phasor variable used in the numerical simulation of dynamic models. The chain rule can be applied to calculate the derivative of this variable:

$$
\begin{aligned}
\frac{d}{dt}\langle x \rangle(t) &= \frac{d}{dt}\left(A(t)\,e^{j\theta(t)}\right) \\
&= \frac{d}{dt}\left(A(t)\,e^{j\varphi(t)-j\omega_c t}\right) \\
&= \frac{d}{dt}\left(A(t)\,e^{j\varphi(t)}\right)e^{-j\omega_c t} + \left(A(t)\,e^{j\varphi(t)}\right)\frac{d}{dt}e^{-j\omega_c t} \\
&= \frac{d}{dt}\left(A(t)\,e^{j\varphi(t)}\right)e^{-j\omega_c t} - j\omega_c\left(A(t)\,e^{j\varphi(t)}\right)e^{-j\omega_c t} \\
&= \left(\frac{d}{dt}x_a(t)\right)e^{-j\omega_c t} - j\omega_c\,x_a(t)\,e^{-j\omega_c t}
\end{aligned}
\qquad (8.13)
$$

Following the same logic as before, we can write

$$\frac{d}{dt}\langle x \rangle(t) = \left\langle \frac{d}{dt}x_a(t) \right\rangle - j\omega_c \langle x \rangle(t) \tag{8.14}$$

If the signal $x_a(t)$ was extracted from a real signal, for example using Fourier Analysis as in [2], it is only an approximation and this derivative equation is also an approximation because it depends on the signal itself and its derivative $\frac{d}{dt}x_a(t)$:

$$\frac{d}{dt}\langle x \rangle(t) \approx \left\langle \frac{d}{dt}x_a(t) \right\rangle - j\omega_c \langle x \rangle(t) \tag{8.15}$$

8.7 Dynamic phasor power system component models

In the following, it is shown how dynamic phasors can be applied to typical components, inductance and capacitance, used in power system simulation.

8.7.1 Inductance model

In order to describe an inductance using dynamic phasors, (8.14) has to be applied to the differential equation (8.16) which describes an inductance:

$$\frac{d}{dt}i_L(t) = \frac{1}{L} \cdot v_L(t) \tag{8.16}$$

This results (8.17) for the fundamental dynamic phasor:

$$\frac{d}{dt}\langle i_L \rangle_1(t) = \frac{1}{L} \cdot \langle v_L \rangle_1(t) - j\omega_s \langle i_L \rangle_1(t) \tag{8.17}$$

8.7.2 Capacitance model

The same approach can used for (6.18) describing a capacitance:

$$i_C(t) = C \cdot \frac{dv_C(t)}{dt} \tag{8.18}$$

The fundamental dynamic phasor is written down in (8.19):

$$\frac{d}{dt}\langle v_C \rangle_1(t) = \frac{1}{C} \cdot \langle i_C \rangle_1(t) + j\omega_s \langle v_C \rangle_1(t) \tag{8.19}$$

8.8 Dynamic phasors and resistive companion models

In commercial electromagnetic transient (EMT) solvers, the nodal analysis method is often employed to solve the system in time domain variables. The same concept can be applied to dynamic phasor models. This has the advantage that high frequency bandpass signals can be shifted toward 0 Hz. Therefore, the maximum frequency of the simulated system can be decreased for many simulation scenarios which enables smaller sampling rates for the simulation. Still, the dynamic behavior as in EMT simulations is maintained.

To support inductances and capacitance with the nodal analysis approach, it is required to derive the resistive companion models. For the derivation of the resistive companion models, the dynamic phasor inductance and capacitance equations presented in the previous section are discretized using the trapezoidal rule which is depicted in (8.20):

$$y(k + 1) = y(k) + \frac{x(k) + x(k + 1)}{2} \cdot \Delta t \tag{8.20}$$

8.8.1   Inductance model

Applying the trapezoidal rule to the fundamental dynamic phasor (8.17) results in (8.21) for time step k:

$$i_L(k + 1) = i_L(k)$$

$$+ \frac{\Delta t}{2} \left[\frac{1}{L} \cdot v_L(k + 1) - j\omega \cdot i_L(k + 1) + \frac{1}{L} \cdot v(k) - j\omega \cdot i_L(k) \right] \tag{8.21}$$

Equation (8.21) can be rewritten as:

$$i_L(k + 1) = \frac{a - jab}{1 + b^2} v_L(k + 1) + \frac{1 - b^2 - j2b}{1 + b^2} i_L(k) + \frac{a - jab}{1 + b^2} v_L(k) \tag{8.22}$$

where

$$a = \frac{\Delta t}{2L} \quad b = \frac{\Delta t \omega}{2}$$

Equation (8.22) shows that the inductor current at step $k + 1$ is dependent on the voltage and current of time step k and the voltage at time step $k + 1$. Therefore, the inductor can be substituted with a resistance R_L in parallel to a current source $A_L(k)$ as depicted in Figure 8.10. The value of R_L is constant for a given frequency

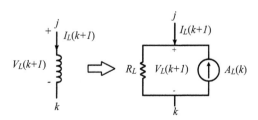

Figure 8.10   Equivalent circuit of an inductor

and time step whereas the value of $A_L(k)$ depends on the previous time step and has to be updated for every step.

The current source and resistance values are calculated according to (8.23):

$$A_L(k) = \frac{1 - b^2 - j2b}{1 + b^2} i_L(k) + \frac{a - jab}{1 + b^2} v_L(k) R_L = \frac{1 + b^2}{a - jab} \qquad (8.23)$$

8.8.2 Capacitance model

Following the approach for the inductance, (8.19) is discretized as in (8.24):

$$v_C(k + 1) = v_C(k)$$
$$+ \frac{\Delta t}{2} \left[\frac{1}{C} \cdot i_C(k + 1) - j\omega \cdot v_C(k + 1) + \frac{1}{C} \cdot i_C(k) - j\omega \cdot v_C(k) \right] \qquad (8.24)$$

The equation can be rewritten as well:

$$i_C(k + 1) = \frac{1 + jb}{a} v_C(k + 1) - \frac{1 - jb}{a} v_C(k) - i_C(k) \qquad (8.25)$$

where

$$a = \frac{\Delta t}{2C} \quad b = \frac{\Delta t \omega}{2}. \qquad (8.26)$$

The capacitor in time step $k + 1$ can also be substituted with a resistance R_C in parallel to a current source $A_C(k)$ as shown in Figure 8.11.

The current source and resistance values are calculated according to (8.27):

$$A_C(k) = \frac{1 - jb}{a} v_C(k) + i_C(k) R_C = \frac{a}{1 + jb} \qquad (8.27)$$

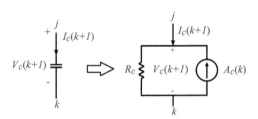

Figure 8.11   Equivalent circuit of a capacitor

8.9   Resistive companion simulation example

The following simulation example demonstrates that dynamic phasor variables are indeed just frequency-shifted compared to their time domain counterparts. In this

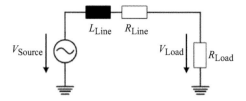

Figure 8.12   Example circuit

case, a simple circuit as depicted in Figure 8.12 is simulated using EMT and dynamic phasor variables.

The circuit consists of an AC voltage source of $V_{Source} = 1$ kV peak voltage with a resistance of $R_{Source} = 1\,\Omega$, an RX-series element of $R_{Line} = 1$ and $L_{Line} = 100$ mH and a load resistance of $R_{Load} = 100\,\Omega$. Internally, the voltage source is transformed to its Norton equivalent. The simulation scenario is as follows: At 0.2 s, the load resistance is decreased to 50 and, at 0.4 s, the frequency of the AC voltage source is decreased from 50 Hz to 45 Hz. This scenario is simulated for a time step of 50 µs which is often used by commercial EMT solvers.

In the following, we compare the voltage V_{Load} across the load resistance. As can be seen in Figure 8.13, the results are almost identical for time steps of 50 µs. Figure 8.13 shows the EMT results, the absolute value of the fundamental dynamic phasor and the time domain signal of the fundamental dynamic phasors after it is shifted back by 50 Hz in the frequency domain. The mean squared error for V_{Load} in the presented time interval is $0.3973 \cdot 10^{-4}$.

The simulation was conducted using the nodal analysis approach that was mentioned in the previous section. The following shows how the system matrix was constructed. First, the system matrix has the dimensions 3×3 but due to the voltage source it is extended by one additional node.

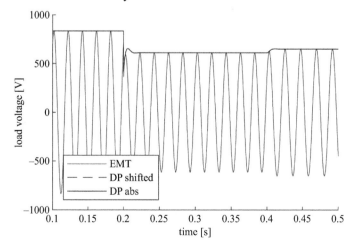

Figure 8.13   EMT and dynamic phasor results

Adding the voltage source results in:

$$
\begin{array}{cccc}
0 & 0 & 0 & 1 \\
0 & 0 & 0 & 0 \\
0 & 0 & 0 & 0 \\
1 & 0 & 0 & 0
\end{array}
$$

Adding the inductance affects four elements because none of the two terminals are connected to the ground node:

$$
\begin{array}{cccc}
0.005 - j7.852 \cdot 10^{-6} & 0.005 - j7.852 \cdot 10^{-6} & 0 & 1 \\
0.005 - j7.852 \cdot 10^{-6} & 0.005 - j7.852 \cdot 10^{-6} & 0 & 0 \\
0 & 0 & 0 & 0 \\
1 & 0 & 0 & 0
\end{array}
$$

Also the line resistor stamps a 2×2 block matrix into the system matrix:

$$
\begin{array}{cccc}
0.005 - j7.852 \cdot 10^{-6} & 0.005 - j7.852 \cdot 10^{-6} & 0 & 1 \\
0.005 - j7.852 \cdot 10^{-6} & 1.005 - j7.852 \cdot 10^{-6} & -1 & 0 \\
0 & -1 & 1 & 0 \\
1 & 0 & 0 & 0
\end{array}
$$

Adding the load resistor only affects one element, since it is connected to ground.

$$
\begin{array}{cccc}
0.005 - j7.852 \cdot 10^{-6} & 0.005 - j7.852 \cdot 10^{-6} & 0 & 1 \\
0.005 - j7.852 \cdot 10^{-6} & 1.005 - j7.852 \cdot 10^{-6} & -1 & 0 \\
0 & -1 & 1.01 & 0 \\
1 & 0 & 0 & 0
\end{array}
$$

The source vector just has the voltage source stamp:

$$
\begin{array}{c}
0 \\
0 \\
0 \\
1000
\end{array}
$$

The solution for the first step is computed by solving the following equation:

$$
\begin{bmatrix}
0.005 - j7.852 \cdot 10^{-6} & 0.005 - j7.852 \cdot 10^{-6} & 0 & 1 \\
0.005 - j7.852 \cdot 10^{-6} & 1.005 - j7.852 \cdot 10^{-6} & -1 & 0 \\
0 & -1 & 1.01 & 0 \\
1 & 0 & 0 & 0
\end{bmatrix}
\begin{bmatrix} x_1 \\ x_2 \\ x_3 \\ x_4 \end{bmatrix}
=
\begin{bmatrix} 0 \\ 0 \\ 0 \\ 1,000 \end{bmatrix}
$$

$$
\begin{bmatrix}
1000 \\
48.061 - j0.718 \\
47.585 - j0.7.115 \\
-0.475 + j0.007
\end{bmatrix}
$$

8.10   Accuracy

Inductors and capacitors are among the most important building blocks of electrical networks and represent two important elements for representing dynamical models: integration and differentiation.

Considering the current as input signal and the voltage as output signal, an inductor represents a differentiation element and a capacitor an integration element. Therefore, it is interesting to analyze the numerical accuracy of these two elements using the trapezoidal rule as discretization method and comparing EMT and DP.

Since an integrator and a differentiator are both linear time-invariant systems, a sinusoidal input signal of the form $e^{j\omega t}$ results in a sinusoidal output signal of the same frequency $Ae^{j\omega t}$, where the complex number A represents the amplitude and phase change. This signal can be used to compare the input–output relation of an ideal differentiation/integration element to its discretized equivalent. This approach is used in [4] to evaluate the model parameter accuracy of inductors and capacitors for the DP model. Here, the accuracy of the output signal is calculated in a similar way.

First, the dynamic phasor model of the capacitor model is expressed in terms of analytical signals using (8.7). Considering i_C as the input and v_C as the output, the capacitor behaves as an integrator:

$$\frac{d}{dt}\langle v_C \rangle = \frac{1}{C} i_C(t)e^{-j\omega_s t} - j\omega_s v_c(t)e^{-j\omega_s t} = e^{-j\omega_s t}\left(\frac{1}{C}i_C(t) - j\omega_s v_C(t)\right)$$

(8.28)

Applying the trapezoidal rule yields

$$\frac{\langle v_C \rangle(k) - \langle v_C \rangle(k-1)}{\Delta t} = e^{-j\omega_s \Delta t}\left[\begin{array}{c} \frac{1}{2C}\left(i_C(k) + i_C(k-1)e^{j\omega_s \Delta t}\right) \\ -j\frac{\omega_s}{2}\left(v_C(k) + v_C(k-1)e^{j\omega_s \Delta t}\right) \end{array}\right].$$

(8.29)

Expressing the remaining phasors as analytical signals results in

$$\langle v_C \rangle(k) - \langle v_C \rangle(k-1)e^{-j\omega_s \Delta t} = \frac{\Delta t}{2C}\left(i_C(k) + i_C(k-1)e^{j\omega_s \Delta t}\right)$$

$$-j\frac{\omega_s \Delta t}{2}\left(v_C(k) + v_C(k-1)e^{j\omega_s \Delta t} v_C(k)\left(1 + j\frac{\omega_s \Delta t}{2}\right)\right)$$

$$= \frac{\Delta t}{2C}\left(i_C(k) + i_C(k-1)e^{j\omega_s \Delta t}\right) + \left(1 - j\frac{\omega_s \Delta t}{2}\right)v_C(k-1)e^{j\omega_s \Delta t} \quad (8.30)$$

This can be shortened using the variables a and b corresponding to the previously derived capacitor model (8.26):

$$v_C(k) = \frac{a}{1+jb}\left(i_C(k) + i_C(k-1)e^{j\omega_s \Delta t}\right) + \frac{1-jb}{1+jb}v_C(k-1)e^{j\omega_s \Delta t} \quad (8.31)$$

Inserting a sinusoidal signal $i_C(k) = e^{j\omega k\Delta t}$ results in a sinusoidal output signal of the same frequency $v_C(k) = Ae^{j\omega k\Delta t}$.

$$Ae^{j\omega_s k\Delta t} = \frac{a}{1+jb}\left(e^{j\omega k\Delta t} + e^{j\omega(k-1)\Delta t}e^{j\omega_s \Delta t} + \frac{1-jb}{1+jb}v_C(k-1)e^{j\omega_s \Delta t}\right) \quad (8.32)$$

To further simplify (8.32), Euler's formula can be applied with $\phi = (\omega_s - \omega)\Delta t$:

$$A = \frac{a\left(1 + e^{j(\omega_s-\omega)\Delta t}\right)}{1+jb - (1-jb)e^{j(\omega_s-\omega)\Delta t}} = \frac{a(1 + \cos(\phi) + \sin(\phi))}{1+jb - (1-jb)(\cos(\phi) + \sin(\phi))}$$

$$A = \frac{a(1+\cos(\phi)+\sin(\phi))(1-\cos(\phi)-b\sin(\phi)+j(-b+\sin(\phi)-b\cos(\phi)))}{(1-\cos(\phi)-b\sin(\phi))^2 + (b-\sin(\phi)+b\cos(\phi))^2}$$

$$= j\frac{a(-b-b\cos(\phi)+\sin(\phi))}{b^2 + b^2\cos(\phi) - 2b\sin(\phi) - \cos(\phi) + 1}$$

$$= j\frac{a(-b-b\cos(\phi)+\sin(\phi))}{-b(-b-b\cos(\phi)+b\sin(\phi)) - b\sin(\phi) - \cos(\phi) + 1}$$

$$= j\frac{a}{-b + \frac{1-b\sin(\phi)-\cos(\phi)}{-b-b\cos(\phi)+\sin(\phi)}}$$

$$\quad (8.33)$$

Using the identity $1 - \cos(\phi) = \tan\left(\frac{\phi}{2}\right)\sin(\phi)$ and substitution of a and b yields

$$A = j\frac{a}{-b + \tan\left(\frac{\phi}{2}\right)} = j\frac{\Delta t/C}{-\Delta t\omega_s + 2\tan((f_s-f)\pi\Delta t)} \quad (8.34)$$

Now, we can calculate the relation between the analytic $A_{a,C} = \frac{1}{j\omega C}$ and the numerical solution $A_{n,C} = A$:

$$\frac{A_{n,C}}{A_{a,C}} = j\omega C \cdot j\frac{\Delta t}{-\omega_s\Delta t + 2\tan((f_s-f)\pi\Delta t)} = \frac{-\omega\Delta t}{-\omega_s\Delta t + 2\tan((f_s-f)\pi\Delta t)}$$

$$= \frac{-\pi f\Delta t}{-\pi f_s\,\Delta t + \tan((f_s-f)\pi\Delta t)}$$

$$\quad (8.35)$$

Hence, the relative error is given by

$$\frac{A_{n,C} - A_{a,C}}{A_{a,C}} = \frac{A_{n,C}}{A_{a,C}} - 1 = \frac{-\pi f\Delta t}{-\pi f_s\,\Delta t + \tan((f_s-f)\pi\Delta t)} - 1 \quad (8.36)$$

Figure 8.14 presents the relative error for the differentiation element and varying shifting frequencies when the input signal frequency is 50 Hz. It is visible that the relative error approaches zero if the shifting frequency is equal to the signal

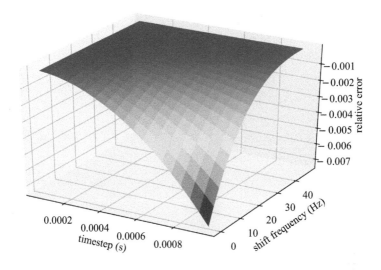

Figure 8.14 Relative error for integrator and 50 Hz signal

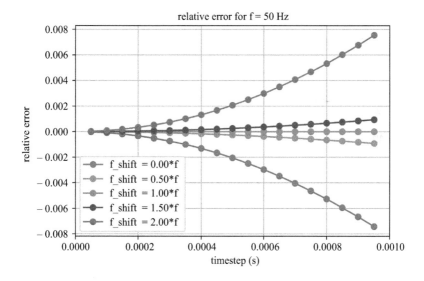

Figure 8.15 Relative error for integrator and 50 Hz signal

frequency even if the integration time step is increased. If the shifting frequency is zero, which represents the EMT simulation, the relative error reaches its maximum in the evaluated interval.

For shifting frequencies higher than the signal frequency, the error grows again as depicted in Figure 8.15. When the signal frequency is increased, the relative error grows faster with respect to the integration time steps even if the relation between shifting and signal frequency is the same in relative terms. This is demonstrated in Figure 8.16 for a signal frequency of 1 kHz.

Similarly, it can be shown that the relative error for a differentiation element is defined by (8.37) since the relation between the two amplitudes is the inverse

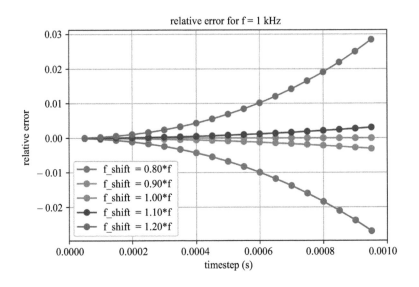

Figure 8.16   Relative error for integrator and 1 kHz signal

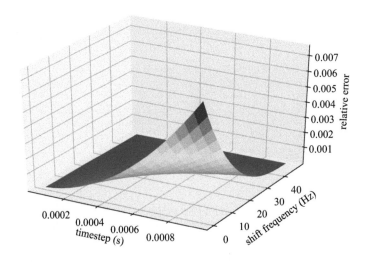

Figure 8.17   Relative error for differentiator and 50 Hz signal

compared to the relation for integration (8.36). The relative error associated to a 50 Hz signal as input to a differentiation element is depicted in Figure 8.17:

$$\frac{A_{n,L} - A_{a,L}}{A_{a,L}} = \frac{A_{n,L}}{A_{a,L}} - 1 = \frac{-\pi f_s \, \Delta t + \tan((f - f_s)\pi \Delta t)}{-\pi f \Delta t} - 1 \tag{8.37}$$

8.11 DP and EMT accuracy simulation example

To demonstrate the effect of frequency shifting for different sampling times, the previous RL circuit as shown in Figure 8.12 is simulated again but this time with different parameters and simulation time steps. The parameters are $V_{Source} = 230$ V peak, $R_{Source} = 1\ \Omega$, $R_{Line} = 1\ \Omega$, $L_{Line} = 20$ mH, and $R_{Load} = 10\ \Omega$. In contrast to the previous simulation, the source frequency is 60 Hz while the shifting frequency for DP is still 50 Hz.

Figure 8.18 depicts the voltage on both sides of the RL line element for dynamic phasors and EMT. Since both simulations are sampled with a small time step of 100 μs, the results after back shifting the DP results look the same. Figure 8.19 shows the phases of the dynamic phasors at the two nodes for the same scenario. To compensate for the difference of 10 Hz between the signal frequency and the shifting frequency, the phases are varying continuously.

Figure 8.20 compares DP and EMT results for a simulation with a time step of 10 ms and again a shifting frequency of 50 Hz for the 60 Hz signal. Even though the

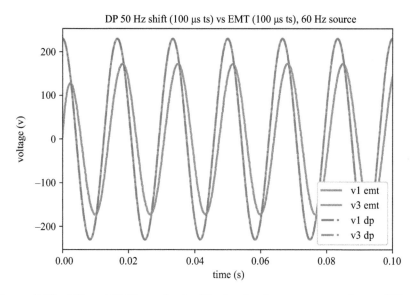

Figure 8.18   DP and EMT results for RL circuit after post processing at 100 μs time step

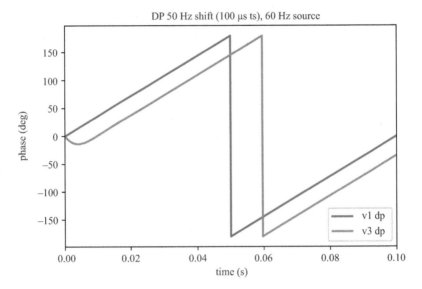

Figure 8.19   DP phases rotating due to difference between shifting and signal frequency

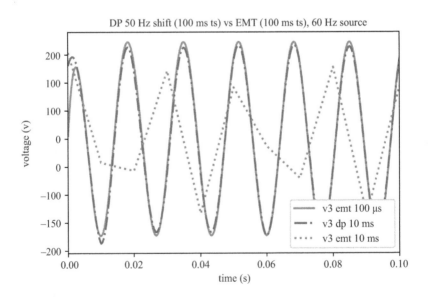

Figure 8.20   DP and EMT results for RL circuit after post processing at 10 ms time step

shifting frequency does not exactly match the signal frequency, the DP simulation error is much smaller, which is line with the discussion in the previous section.

8.12 Summary

This section introduces the dynamic phasor concept which is related to e.g. complex envelope in electronics and communication. It is explained how dynamic phasors can be applied to basic resistive companion models of inductors and capacitors. The first simulation example presents the combination of dynamic phasors with nodal analysis and resistive companion models.

The accuracy of dynamic phasors is quantified for dynamic models using the trapezoidal method for discretization. Finally, an example simulation demonstrates the advantages of the dynamic phasor approach compared to traditional EMT simulation for larger simulation time steps.

Exercise 1 In this exercise, you are supposed to build a small RL-circuit as depicted in Figure 8.21 using dynamic phasor component models and the Modelica language. All component models are given except for the inductor. After completing the implementation of the inductor model, explain the transient behavior of the circuit and write additional equations to transform the voltage and current phasor variables to physical variables by shifting the results by 50 Hz.

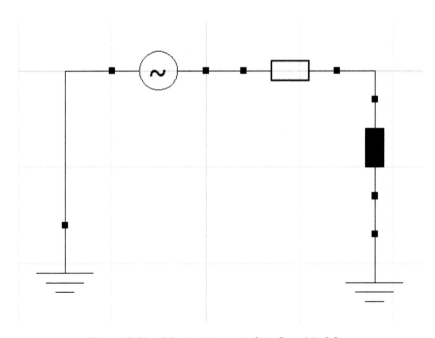

Figure 8.21   RL-circuit created in OpenModelica

First we need to define a connector:

```
within ModelicaExercises.DynPhasor.Interfaces;
connector Pin "Pin with voltage potential and current flow"
 Modelica.SIunits.ComplexVoltage v "Complex potential at
the pin";
 flow Modelica.SIunits.ComplexCurrent i "Complex current
flowing into the pin";
end Pin;
```

Using this connector, we can create models for a voltage source

```
within ModelicaExercises.DynPhasor;
model VoltageSource
 parameter Modelica.SIunits.Voltage Vref = 10 "Voltage
reference amplitude";
 parameter Modelica.SIunits.Angle phiVref = 0 "Initial
voltage reference angle";
 Interfaces.Pin Pin1;
 Interfaces.Pin Pin2;
 Modelica.SIunits.ComplexVoltage v "Voltage between Pin1
and Pin2";
 Modelica.SIunits.ComplexCurrent i "Current flowing from
Pin1 to Pin2";
 Modelica.SIunits.ComplexVoltage v1 "Voltage at Pin1";
 Modelica.SIunits.ComplexVoltage v2 "Voltage at Pin2";
equation
 v = v1 - v2;
 v1 = Pin1.v;
 v2 = Pin2.v;
 i = Pin1.i;
 i = -Pin2.i;
 v = Complex(Vref*cos(phiVref), Vref*sin(phiVref));
end VoltageSource
```

a ground node

```
within ModelicaExercises.DynPhasor;
model Ground
 Interfaces.Pin Pin1;
 Modelica.SIunits.ComplexVoltage v;
equation
 v = Pin1.v;
 v = Complex(0,0);
end Ground;
```

and a resistor

```
within ModelicaExercises.DynPhasor;
model Resistor
 parameter Modelica.SIunits.Resistance R = 1;
 Interfaces.Pin Pin1;
 Interfaces.Pin Pin2;
 Modelica.SIunits.ComplexVoltage v "Voltage between Pin1
and Pin2";
```

```
 Modelica.SIunits.ComplexCurrent i "Current flowing from
Pin1 to Pin2";
 Modelica.SIunits.ComplexVoltage v1 "Voltage at Pin1";
 Modelica.SIunits.ComplexVoltage v2 "Voltage at Pin2";
 Modelica.SIunits.Voltage V "Voltage amplitude";
 Modelica.SIunits.Current I "Current amplitude";
equation
 v = v1 - v2;
 v1 = Pin1.v;
 v2 = Pin2.v;
 i = Pin1.i;
 i = -Pin2.i;
 V = 'abs'(v);
 I = 'abs'(i);
 v = R*i;
end Resistor;
```

For the inductor, the equation block is left to the reader:

```
within ModelicaExercises.DynPhasor;
model Inductor
 parameter Modelica.SIunits.Inductance L = 1e-3;
 parameter Modelica.SIunits.Frequency fnom = 50;
 Modelica.SIunits.Reactance X = 2*pi*fnom * L;
 Interfaces.Pin Pin1;
 Interfaces.Pin Pin2;
 Modelica.SIunits.ComplexVoltage v "Voltage between Pin1
and Pin2";
 Modelica.SIunits.ComplexCurrent i "Current flowing from
Pin1 to Pin2";
 Modelica.SIunits.ComplexVoltage v1 "Voltage at Pin1";
 Modelica.SIunits.ComplexVoltage v2 "Voltage at Pin2";
 Modelica.SIunits.Voltage V "Voltage amplitude";
 Modelica.SIunits.Current I "Current amplitude";
equation
end Inductor;
```

The circuit in components is defined as follows:

```
within ModelicaExercises.DynPhasor.Examples;
model VoltageSource_RL
 ModelicaExercises.DynPhasor.VoltageSource voltage-
Source;
 ModelicaExercises.DynPhasor.Resistor resistor(R=1);
 ModelicaExercises.DynPhasor.Ground ground;
 ModelicaExercises.DynPhasor.Inductor inductor(L=0.001);
 ModelicaExercises.DynPhasor.Ground ground1;
equation
 connect(inductor.Pin1, ground1.Pin1);
 connect(resistor.Pin1, inductor.Pin2);
 connect(ground.Pin1, voltageSource.Pin1);
 connect(voltageSource.Pin2, resistor.Pin2);
```

```
end VoltageSource_RL;
```

Solution
The equation block of the inductor looks like this.

```
equation
  v.re = L*der(i.re) - X*i.im;
  v.im = L*der(i.im) + X*i.re;
```

Exercise 2 In this exercise, a circuit with the following parameters is considered:

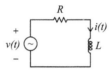

Figure 8.22   RL circuit

$$R = 0.5 \ \Omega, L = 10 \ \text{mH}, v(t) = 10 \sin(\omega t), f = 50 \ \text{Hz}, i(0) = 0 \ \text{A}$$

The step size should be $\Delta t = 1\text{ms}$.

1. Draw the equivalent circuit using dynamic phasors and the resistive companion method. For the discretization, apply the trapezoidal integration rule.
2. Starting from the equivalent circuit of task 1, calculate the conduction matrix and source vector and solve the circuit for the first time step, $i(t = 1\text{ms})$, using nodal analysis.

Solution
Draw the equivalent circuit.

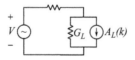

Figure 8.23   Equivalent circuit

$$V = 0 - j10$$

$$G_L = \frac{1}{\frac{2L}{\Delta t} + j\omega L}$$

$$A_L(k) = \frac{1}{\frac{2L}{\Delta t} + j\omega L} V_L(k) + \frac{\frac{2}{\Delta t} - j\omega}{\frac{2}{\Delta t} + j\omega} I_L(k)$$

$$I_L(0) = 0$$

Write the admittance matrix

$$\begin{bmatrix} \dfrac{1}{R} & -\dfrac{1}{R} & 1 \\ -\dfrac{1}{R} & \dfrac{1}{R}+G_L & 0 \\ 1 & 0 & 0 \end{bmatrix}$$

and the source vector.

$$\begin{bmatrix} 0 \\ -A_L(0) \\ V \end{bmatrix}$$

Calculate the first step from the following system:

$$\begin{bmatrix} E_1(1) \\ E_2(1) \\ I_1(1) \end{bmatrix} = \begin{bmatrix} \dfrac{1}{R} & -\dfrac{1}{R} & 1 \\ -\dfrac{1}{R} & \dfrac{1}{R}+G_L & 0 \\ 1 & 0 & 0 \end{bmatrix}^{-1} \begin{bmatrix} 0 \\ -A_L(0) \\ V \end{bmatrix}$$

where

$$A_L(0) = \frac{-j10}{\frac{2L}{\Delta t}+j\omega L}$$

The result for $i(t = 1 \text{ ms})$ is:

$$\begin{bmatrix} 0.073 - j9.523 \\ -j10 \\ 0.146 + j0.953 \end{bmatrix}$$

References

[1] P. Kundur, N. J. Balu and M.G. Lauby, *Power System Stability and Control,* Vol. 7. New York, NY: McGraw-Hill, 1994.
[2] S. R. Sanders, J. M. Noworolski, X. Z. Liu, *et al.*, "Generalized averaging method for power conversion circuits," *IEEE Transactions on Power Electronics*, vol. 6, no. 2, pp. 251–259, 1991.
[3] V. Venkatasubramanian, H. Schattler and J. Zaborszky, "Fast time varying phasor analysis in the balanced three-phase large electric power system," *IEEE Transactions on Automatic Control*, vol. 40, no. 11, pp. 1975–1982, 1995.
[4] K. Strunz, R. Shintaku and F. Gao, "Frequency-adaptive network modeling for integrative simulation of natural and envelope waveforms in power

systems and circuits," *IEEE Transactions on Circuits and Systems I: Regular Papers*, vol. 53, no. 12, pp. 2788–2803, 2006.

[5]   T. Demiray, G. Andersson and L. Busarello, "Evaluation study for the simulation of power system transients using dynamic phasor models," in Transmission and Distribution Conference and Exposition: Latin America, 2008 IEEE/PES. IEEE, 2008, pp. 1–6.

[6]   P. Mattavelli, A. M. Stankovic and G. C. Verghese, "SSR analysis with dynamic phasor model of thyristor-controlled series capacitor," *IEEE Transactions on Power Systems*, vol. 14, no. 1, pp. 200–208, 1999.

[7]   P. C. Stefanov and A. M. Stankovic, "Modeling of UPFC operation under unbalanced conditions with dynamic phasors," *IEEE Transactions on Power Systems*, vol. 17, no. 2, pp. 395–403, 2002.

[8]   A. M. Stankovic and T. Aydin, "Analysis of asymmetrical faults in power systems using dynamic phasors," *IEEE Transactions on Power Systems*, vol. 15, no. 3, pp. 1062–1068, 2000.

[9]   R. H. Salim and R. A. Ramos, "A model-based approach for small-signal stability assessment of unbalanced power systems," *IEEE Transactions on Power Systems*, vol. 27, no. 4, pp. 2006–2014, 2012.

[10] S. Chandrasekar and R. Gokaraju, "Dynamic phasor modeling of type 3 dfig wind generators (including SSCI phenomenon) for short-circuit calculations," *IEEE Transactions on Power Delivery*, vol. 30, no. 2, pp. 887–897, 2015.

[11] T. Demiray, F. Milano and G. Andersson, "Dynamic phasor modeling of the doubly-fed induction generator under unbalanced conditions," in *Power Tech, 2007 IEEE Lausanne*. IEEE, 2007, pp. 1049–1054.

[12] A. M. Stankovic, S. R. Sanders and T. Aydin, "Dynamic phasors in modeling and analysis of unbalanced polyphase ac machines," *IEEE Transactions on Energy Conversion*, vol. 17, no. 1, pp. 107–113, 2002.

[13] P. Zhang, J. R. Marti and H. W. Dommel, "Synchronous machine modeling based on shifted frequency analysis," *IEEE Transactions on Power Systems*, vol. 22, no. 3, pp. 1139–1147, 2007.

[14] S. A. Maas, *Nonlinear Microwave and RF Circuits*. London: Artech House, 2003.

[15] A. Suárez. *Analysis and Design of Autonomous Microwave Circuits*, Vol. 190. New York, NY: John Wiley & Sons, 2009.

[16] H. Simon, *Communication Systems*, 5th ed. New York, NY: Wiley Publishing, 2009.

[17] J. G. Proakis and M. Salehi, *Communication Systems Engineering*, Vol. 2. Prentice, NJ: Prentice Hall, 1994.

[18] P. J. Schreier and L. L. Scharf, *Statistical Signal Processing of Complex-Valued Data: The Theory of Improper and Noncircular Signals*. Cambridge: Cambridge University Press, 2010.

[19] E. Ngoya and R. Larcheveque, "Envelope transient analysis: a new method for the transient and steady-state analysis of microwave communication circuits and systems," in IEEE Microwave Theory and Techniques Symposium, 1996, pp. 1365–1368.

Chapter 9

Modeling of converters as switching circuits

Ferdinanda Ponci[1,2] and Antonello Monti[1,2]

9.1 Simulation of power electronics systems

Power electronics are mostly developed as technology supporting the design of efficient power supplies. Through the years, the range of applications of power electronics has been dramatically growing.

One of the key developments has been the application in electrical drives that brought power electronics into the industry. As first stage, power electronics made possible the application of electrical drives in DC and then the transition to AC drives.

In all these cases, the focus of the power electronic simulation has been on the single component development. In such a scenario, the main challenge was the interaction among the electronic converter and the electromechanical part of the system.

For large drives, the main time constant for the electrical machine can be in the range of seconds, while switching frequency of devices is currently in the kHz range.

This simple example already clarifies that there is a wide spread of eigenvalues making the simulation difficult to set-up.

It appeared then historically clear from the beginning that power electronic systems require always a variety of models from the device level to the system level.

In simple sense, we can define the following categories:

- Device level modeling starting from the physics of semiconductors
- Switching level modeling approximating semiconductors with idea switches
- Averaging methods substituting switching mode power supplies with linear amplifiers

Clearly the bigger the system, the more critical is the selection of the appropriate model. To illustrate the concept, the system in Figure 9.1 can be a good reference. This picture represents a simplified schematic of a modern ship based on medium voltage distribution feeders.

The adoption of DC in ships has been brought by the need to increase the overall efficiency and the reducing fuel consumption. This transformation is a

[1]Institute for Automation of Complex Power Systems, RWTH Aachen University, Germany
[2]Fraunhofer FIT Center for Digital Energy, Germany

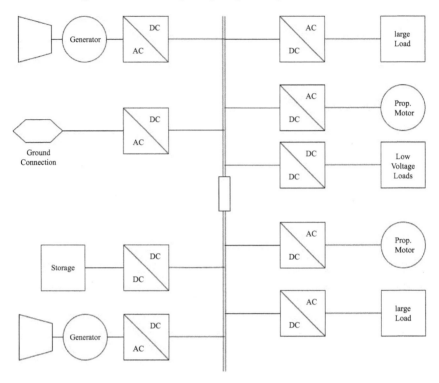

Figure 9.1   A modern DC-based ship system

subsequent transformation after the moving in the 1990s towards electrical propulsion. The more modern way is the so called "All Electric Ship" in which every
actuator is connected to the electrical infrastructure.

Such type of ship is a very good example of power electronics power systems.
Here the modeling challenges to move from components to systems are even more
evident than in electrical drive because we are dealing with the interaction of a set
of converters.

Modeling such system is then particularly difficult because new challenges such
as interaction of controllers in the same bus is a key topic that averaging model may
not address properly. We reach then a situation in which ultra-fast simulation
methods become critical to keep the simulation time under reasonable limits.

While ship systems have been the first application in which the challenge of
large power electronic systems emerged, the growing presence of renewables is
bringing this modeling challenge also to terrestrial systems. New scenarios of
application become then reasonable and possible. Figure 9.2 shows a possible
structure for a DC microgrid.

The consideration that most of modern devices need a DC bus at some point of
the energy processing is reopening the competition between DC and AC for the
development of future energy grids.

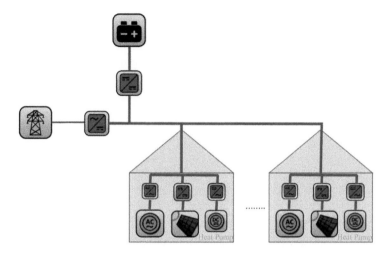

Figure 9.2   A structure for a DC microgrid

DC microgrids are an example of local energy system in which customers in low voltage are clustered by means of a DC bus.

Here again, as in the ship case, we experience the interaction of a large set of power electronic devices some of them working as sources and other as loads.

While the growing the available computational power is making possible to consider the simulation based on switching models, still when the number of final users grows to a large city quarter, the application of averaging techniques supporting large time steps become a favorite solution.

All in all, power electronic systems are becoming more and more important. The need of a variety of modeling approaches for the converter devices is then of growing importance to study questions ranging from the efficiency of a single device to the stability of a full bus with several interconnecting devices.

9.2   Role of power electronics in power systems

Power electronics is playing a significant role in modern power systems. Power electronics is one of the main technology drivers behind the evolutions both in energy systems and in transportation systems. Basically, every kind of renewable energy source is connected to the grid by means of a power electronic interface.

On the other hand, loads are becoming smart, which entails also that they are equipped with power electronic converters. The combination of generation, storage, and load made controllable and flexible by power electronics interfaces enables the creation of local energy systems, microgrids and the opportunity to operate islanded portions of the distribution systems as autarkic microgrid. These scenarios are open for a variety of innovations in the technology and operation of power systems, like for example neighborhood energy trading, energy packet distribution strategies, bottom-up blackout recovery.

Other systems too are incrementally more electrified, such as ships or aircrafts, and whereas trains have always been predominantly electric, the electric cars are expected to play a key role in the future of mobility. The across-sectors energy flexibility exchange could maximize renewable energy usage in unprecedented ways.

In the modern power system scenario sketched out above, the analysis, design, verification, validation and certification of infrastructure, control automation, services, regulation, business must rely on models. Hence, the right approach to incorporate power converters in the grid model is key for reliable results.

The selection of the power converter model depends on the modeling purpose and thus eventually on the application and use of the model. In a nutshell, the converter can be represented on many static and dynamic levels of detail, from the ideal source level (simplest) to the full model including model of the junction and parasitic components (most complex), even in a multi-physic form that includes non-electrical phenomena e.g. the thermal exchange. The strategy "the more (detailed) the better" does not work. In fact, more detail means more parameters to set (which may not be known yet or at all by the user), possibly more interfaces, and more (and longer) calculations, with possibly no benefit, if the rest of the system model is not equally detailed or if the application does not need it.

Consider for example the case of the analysis of the power flows on board a fully electric ship, for the purpose of sizing equipment or to determine the impact of large pulsed loads such as electric aircraft catapults. In this case, the power converters may be incorporated as averaged models, capable of representing the power flow control capacity but lacking the switching effects and the topological details of the converter.

A similar case, however, requiring a more detailed model, is the analysis of optimal power flow strategies in a local energy system, microgrid or islanded portion of a distribution system. In this case, a group of converters operate to support the voltage and to guarantee that the demand is met. The power flow is one part of the story; however, the power/voltage controllers must be included and the stability analysis, and the analysis of different dynamic conditions may require a switching model representation of the converter. Even more so in case power quality requirements are to be met, especially if they refer to harmonic injections of the power converter or their active suppression.

If the model of the converter is to be used e.g. for the design of the snubbers, then the model of the junction of the individual switches must be incorporated, together with possible parasitic elements. Similarly consider the case of switching losses affecting the assessment and analysis of efficiency and heat disposal. In particular, in this case, the thermal model of the heat sink, and possibly of more complex thermal infrastructures, must be coupled with the model of the power converter, posing the challenge of joint modelling and simulation of systems with time constants that are orders of magnitude apart.

9.3 Modelling and simulation of power electronics in power systems

Simulation of power electronics is a challenging task. The model can describe the devices at very different level of details. All in all, we can distinguish three main levels:

- *Device level simulation*: in this case, the simulation deals with the physics of the semiconductor and the model tries to capture the internal processes of the semiconductor devices.
- *Switching level simulation*: in this case, the simulation adopts ideal switches. The model of these switches can be different but the principle is always the commutation among a set of configurations.
- *Average simulation*: in this case, the simulation adopts an ideal representation of the converter in which the switching action is not visible and rather the focus is on the main flow of power.

The three levels are of decreasing complexity allowing at the same time an increase in the time step and a corresponding reduction in the computational effort. In this chapter, we review the main theoretical contributions in the area of modelling in reference only the two higher levels. The reason is that typically it is of no use for system level analysis and simulation to consider the physics of the device. It is interesting to observe that, in general, the modeling approaches presented here still fall into two main categories of models introduced at the beginning of the book and, i.e., topological based or state equation based.

As result, there is no need for a specific simulation approach for power electronics in power systems other than these two. Rather, the issue to be addresses lays in the choice of the right modelling approach that fits a given network solver.

9.4 Converter models

This chapter presents a summary of the models of switching circuits. Switching circuits are designed to operate in nominal conditions. However, in practice, deviations from this operating point may occur, because of variations of the source or of the load, perturbations of the switching instant and turning on and off of the entire system. These events are compensated for by the control, which must also be able to change operating condition at will. Consider for example the case of a power supply, whose output must rigorously be kept at a given value, in spite of the unforeseeable variations of the input or load variations, which cause current and thus voltage variations as a consequence. Another example is the motor drive performing torque control in rotating machines. To anticipate and quantify the effects of external variations onto the output of the converter and to design the control, a suitable dynamic model has to be synthesized. The following sections present averaged models, directly derived from the lumped parameter equivalent circuit models, the state space models with their averaged versions and the discrete models.

9.5   Averaged models

The theoretical justification of the derivation of the averaged models lays in the linearity of the averaging operator. As a consequence, Kirchhoff laws and Telleghen theorem, which hold for the instantaneous variables of switching circuits, also hold for the averaged variables. As a consequence, a given switching circuit can be mapped onto a corresponding averaged circuit. These models are frequently adopted for rectifiers and dc/dc converters. Two approaches are possible: the averaging of the entire switching circuit or the averaging of the individual switching component. In this mapping, the linear, time-invariant circuit components remain the same, whereas the ideal switching components are turned into elements subject to simultaneous voltage and current.

The averaged variable, voltage or current, is the average local value of a given variable $x(t)$, defined as:

$$\bar{x}(t) = \frac{1}{T} \int_{t-T}^{t} x(\tau) d\tau \tag{9.1}$$

where T is a conveniently fixed period, typically much smaller than the smallest time constant of the system, in the case of dynamic analysis.

Because we focus on switching circuits, T can be chosen to be the switching period, or better the smallest of the switching periods. The case in which the variables are themselves periodic is particularly interesting because, by choosing T to be equal to the period of variable $x(t)$, the average value of $x(t)$ is then constant.

9.6   Averaged circuits

In these circuits, the variables, originally represented by their instantaneous values, are now replaced by their average values. This implies that also components that were originally nonlinear or time-varying, hence that would change their behavior during the averaging period, must be replaced.

Let us first consider the case of a given circuit composed only of linear, time-invariant elements and ideal switches. The only averaged elements are the switching elements, which are transformed into elements with non-zero terminal voltage and current (instead of alternatively open or short circuits). The new averaged circuit model can then be studied with the methods of network analysis. The averaged circuit model corresponds to the averaged equations of the model.

Let us consider now a simple case of averaged circuit: a motor drive for dc motors comprising a controlled rectifier with sinusoidal source v_{ac} of amplitude V_0, modeled with an ideal voltage source, and linear, time-invariant components. The set source+converter can be modeled as an ideal voltage source whose value is the average value of the output of the converter (Figure 9.3).

A convenient choice of T, the averaging period, is the output period of the converter, corresponding to half the period of v_{ac}. The value of the ideal constant

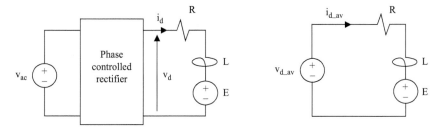

*Figure 9.3   Reference circuit of motor drive converter with instantaneous values
(left) and averages circuit with mean values of the variables (right)*

voltage source v_{d_av} depends on the source v_{ac} and on the turn-on delay of the
switching elements, α. Finally:

$$v_{d\_av} = \frac{2V_0}{\pi} \cos \alpha \tag{9.2}$$

The average circuit thus obtained is determined for the nominal operating
condition. Any variation of the sinusoidal source, or of the turn-on time of the
switching components, causes a variation of the value of the equivalent generator. From the point of view of the load, this results in a step variation of the
supply, resulting, for example, in the case of an inductive load, as an exponential
transient.

This model is simple as the switching reactance is neglected. The averaged
circuit allows the determination of the effect on the load of variations of the
operating conditions, and thus supports the design of a simple feedback controller
to regulate e.g. the load current (and hence the torque in this example).

In summary, the averaged model loses the information on the ripple of the
variables caused by the switching, but it allows the determination of the static
regulation and the transient response of the system.

Let us consider now the case of converters with constant input sources, such as
dc/dc buck and boost converters in various configurations, and pulse width modulation (PWM) inverters. These converters have a constant (or approximatively so)
input voltage V_{in} and linear load, and produce an output voltage V_{out}, which can be
expressed as the product of the source voltage multiplied by a modulating function
$q(t)$, i.e. $V_{out} = V_{in} \cdot q(t)$, where $q(t)$ is the switching function.

To replace the entire source–converter system with one single generator
representing the averaged output of the converter, we average V_{out}, that is, we
average the switching function. This operation is simple thanks the typical
expression of the switching function, whose values belong to a finite set, e.g. 1 and
0 or -1, 1 and 0. The averaging period corresponds to the switching period.

In the particular case in which the switching function is periodic with period
equal to the switching period, the output of the converter is constant.

9.7   Averaged switching elements

With this approach, the switching element is replaced with an averaged circuit. This approach is particularly convenient for circuits with high switching frequency or PWM converters. In this case, the averaging is performed over the switching period under the assumption that the circuit variables are well approximated by their average value and that their variation is slower that the switching period. It is also assumed that the circuit is operating in continuous conduction mode (CCM), hence, is linear.

Under these assumptions, the switching element is characterized through its constraints of the average variables at its terminals. This averaging of the switching components has no effect on the other circuit components, which are anyway assumed to be linear and time-invariant (LTI). The process starts with the basic switching cell and the replacement of its terminal variables with their average values. This yields a representation of the basic cell in terms of a network of passive components and controlled sources, or alternatively an ideal transformer.

Let us consider the case of the buck converter in Figure 9.4. To analyze the switching element, we refer to the basic cell in the configuration of direct converter in Figure 9.5. The input capacitance is not represented here under the assumption that the dc source is fully capable of maintaining a constant dc voltage. Let us suppose that this buck converter is operating with fixed duty cycle "d" and that the

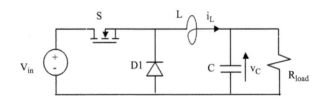

Figure 9.4   Buck converter

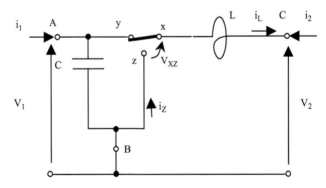

Figure 9.5   Reference switching cell of the direct converter

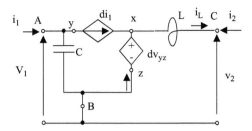

Figure 9.6   Averaged model, based on controlled sources, of the reference switching cell

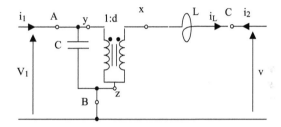

Figure 9.7   Averaged model of the reference switching cell based on the ideal transformer

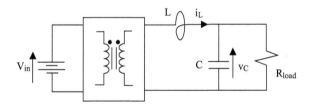

Figure 9.8   Averaged circuit of the buck converter

inductance is such that the current is held constant, so that no large oscillations may occur. The averaged cell model may be represented as a network of passive components and controlled sources as in Figure 9.6, or via passive components and ideal transformer, as in Figure 9.7.

These manipulations and either averaged model of the switching cell yield the averaged circuit of the buck converter shown in Figure 9.8.

Starting from this averaged circuit, it is possible to express the model of the system in terms of state equations of the entire system or to linearize it.

9.7.1   Linearization

The models obtained with the procedure described above are typically nonlinear, and the same can be said for feedback controlled systems. The nonlinearity appears for example, for the presence of a variable that is the product of a state variable time the duty cycle. For nonlinear systems, the control design and the stability analysis are more difficult. It is therefore convenient to linearize such models, thus producing linear models that well represent the behavior of the system in presence of small perturbations and small deviations with respect to the reference condition of the linearization. Such linearized models are thus termed small signal models. To linearize a model, first a nominal condition, around which the linearized representation will hold, must be determined. This is typically a steady-state condition where the switching function is periodic.

The following process represents the linearization in the continuous-time domain. Voltages and currents are replaced with their deviations from the chosen nominal value. Kirchhoff laws still hold for the new circuit with variables representing the deviations. The nonlinear elements, whose characteristic may be derived via the method of the averaged circuit, are replaced with their linearized model obtained with a truncated Taylor expansion of their nonlinear characteristic.

The linear, time-invariant components stay the same. The resulting circuit model represents the relations between the deviations from the nominal condition. This method linearizes the average model, which expresses the whole switching circuit.

An alternative is instead the linearization of the average switching element. As seen before the individual switching cell can be expressed as an average model in terms of controlled sources or an ideal transformer. The application of the linearization method to these cell models yields the small signal model of the switching element.

9.7.2 Considerations on the averaged models

The averaged models are widely used in practice. However, not always the objective of the analysis of converters is the average behavior of the waveforms. Furthermore, the path to the formulation of the average model may include various approximations, which may compromise the validity of the resulting model and the range of applicability. Different operating conditions correspond to different linearized models, with the consequence that the performance of controllers designed based on one linearized models, is not guaranteed in case of significant deviations from the linearization operating points. Finally, the averaged models may be unsuitable to design or describe the behavior for digital controllers.

9.8 State-space models

State-space models are more general models that overcome the shortcomings of average models as discussed above.

State-space models form a very broad group that encompasses average models as particular cases. State-space models comprise continuous and discrete time domain models, and they can be averaged and linearized. In fact, linearized models comprising only linear, time invariant models and piecewise linear models are a very relevant case.

Another very important characteristic of state-space models is that they can be automatically derived from the circuit description, in case of linear, time-invariant circuits.

The state-space models are formulated in terms of state variables, inputs i.e. those external signals that supply or control the system, including disturbances, and the outputs, which are the variables of interest.

Typically, inductor current or flux and capacitance voltage or charge are chosen to be the state variables. The general expressions of the continuous and discrete time state-space models are presented herein. Such equations can be expressed in the more compact matrix form.

9.8.1 Continuous time models

In the simplest case, with no time varying, nonlinear or mutually coupled inductors, the model derivation is as follows. The state-space equations express the voltage of the inductors and the current of the capacitors as functions of the state variables and the inputs. The general expression is:

$$\frac{dx_1}{dt} = \dot{x}_1(t) = f_1(x_1(t), x_2(t), \ldots, u_1(t), \ldots, u_m(t), t)$$

$$\frac{dx_2}{dt} = \dot{x}_2(t) = f_2(x_1(t), x_2(t), \ldots, u_1(t), \ldots, u_m(t), t)$$

$$\vdots$$

$$\frac{dx_n}{dt} = \dot{x}_n(t) = f_n(x_1(t), x_2(t), \ldots, u_1(t), \ldots, u_m(t), t)$$

(9.3)

The state-space model thus obtained can then be averaged with the same method described in the previous sections, however, its final expression is not as readable as the one obtained directly from the averaged circuit.

Consider for example the case of the Čuk converter in Figure 9.9.

This converter can be modeled via the state variables inductor currents i_{L1} and i_{L2}, and capacitor voltage v_c. Following the procedure above, we derive the expression of the inductor voltages and capacitor currents as functions of the state variables, the inputs and the control variables. We assume that the circuit is

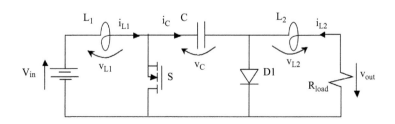

Figure 9.9   Equivalent circuit of the Čuk converter

operating in CCM and we define as $q_1(t)$ and $q_2(t)$ the switching functions of the switching elements S and D1 for which, under our assumptions, the following holds: $q_1(t)+q_2(t)=1$. The equations are expressed in terms of switching functions, which modulate the source and represent the switching of the circuit.

The modeling process yields the following equations that refer to the instantaneous variables, whereas omitting the time variable, for simplicity:

$$\begin{cases} v_{L1} = V_{in}q_1 + (V_{in} - v_c)q_2 \\ i_C = q_1(-i_{L2}) + q_2 i_{L1} \\ v_{L2} = q_2(-v_{out}) + q_1(-v_{out} + v_C) \end{cases} \tag{9.4}$$

Rearranging the equations, considering that $v_{out} = R_{load} \cdot i_{L2}$ and applying the derivative with respect to time, yields:

$$L_1 \frac{di_{L1}}{dt} = V_{in} - q_2 v_C$$

$$C \frac{dv_C}{dt} = -q_1 i_{L2} + q_2 i_{L1} \tag{9.5}$$

$$L_2 \frac{dv_{L2}}{dt} = q_1 v_C - i_{L2} R_{load}$$

It is now possible to average the state space model thus obtained. For this purpose, we have to assume that the variables are approximately constant over the period of averaging, a condition for the value of the variable and its average value to adequately match.

Furthermore, assuming that the switching functions can only take values 0, 1, and -1, yields:

$$\overline{q(t)x(t)} \approx \bar{q}(t)\bar{x}(t) = D\bar{x}(t) \tag{9.6}$$

D being the average value of the switching function, that is, the duty cycle

In this particular case D_1, for switching function q_1, and $(1-D_1)$ for switching function q_2 (given the dependency between q_1 and q_2) appear in the equations:

$$L_1 \frac{d\bar{i}_{L1}}{dt} = V_{in} - (1 - D)\bar{v}_C$$

$$C \frac{d\bar{v}_C}{dt} = -D_1 \bar{i}_{L2} + (1 - D)\bar{i}_{L1} \tag{9.7}$$

$$L_2 \frac{d\bar{v}_{L2}}{dt} = D_1 \bar{v}_C - \bar{i}_{L2} R_{load}$$

The averaged state space model can be represented in the matrix form. To illustrate this formalism, consider the boost converter in Figure 9.10, which is also used as example of linearization.

Let us suppose that this converter is operating in CCM, that $q_1(t)$ is the switching function of S, $q_2(t)$ is the switching function of D and thus in CCM $q_1(t)+q_2(t)=1$. Notice also that $i_{load}=v_C/R$. Inductor current i_L and capacitor voltage v_c are chosen as state variables.

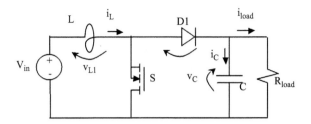

Figure 9.10 Reference equivalent circuit of the boost converter

Omitting the time dependence, the equations in matrix form of the state-space model can be derived for each of the two possible configurations, that is "S on and D off" for the duration $T_{on} = D_1 T$ and "S off and D on" for the duration $T_{off} = D_2 T$.

$$\dot{x} = A_1 x + B_1 u \quad S \quad on$$
$$\dot{x} = A_2 x + B_2 u \quad D \quad on \tag{9.8}$$

where x is a two-element vector, with elements i_L and v_c, and u is the vector of inputs.

In the particular case of the boost converter example adopted here, the equations are:

$$\begin{bmatrix} \dot{i}_L \\ \dot{v}_C \end{bmatrix} = \begin{bmatrix} 0 & 0 \\ 0 & -\dfrac{1}{CR_{load}} \end{bmatrix} \begin{bmatrix} i_L \\ v_C \end{bmatrix} + \begin{bmatrix} \dfrac{V_{in}}{L} \\ 0 \end{bmatrix} \quad S \quad on$$

$$\begin{bmatrix} \dot{i}_L \\ \dot{v}_C \end{bmatrix} = \begin{bmatrix} 0 & -\dfrac{1}{L} \\ \dfrac{1}{C} & -\dfrac{1}{CR_{load}} \end{bmatrix} \begin{bmatrix} i_L \\ v_C \end{bmatrix} + \begin{bmatrix} \dfrac{V_{in}}{L} \\ 0 \end{bmatrix} \quad D \quad on \tag{9.9}$$

and hence in terms of switching functions, the system model is:

$$L\frac{di_L}{dt} = V_{in} - q_2 v_C$$
$$C\frac{dv_C}{dt} = q_2 i_L - \frac{v_C}{R_{load}} \tag{9.10}$$

Which, in the matrix form, is:

$$\begin{bmatrix} \dfrac{di_L}{dt} \\ \dfrac{dv_C}{dt} \end{bmatrix} = \begin{bmatrix} 0 & -\dfrac{q_2}{L} \\ \dfrac{q_2}{C} & -\dfrac{1}{CR_{load}} \end{bmatrix} \begin{bmatrix} i_L \\ v_C \end{bmatrix} + \begin{bmatrix} \dfrac{V_{in}}{L} \\ 0 \end{bmatrix} \tag{9.11}$$

The averaged model can then be expressed as:

$$\bar{A} = A_1 D_1 + A_2 D_2$$
$$\bar{B}\bar{u} = B_1 u D_1 + B_2 u D_2 \tag{9.12}$$

hence in the matrix form:

$$
\begin{bmatrix} \dfrac{d\bar{i}_L}{dt} \\ \dfrac{d\bar{v}_C}{dt} \end{bmatrix} = \begin{bmatrix} 0 & -\dfrac{D_2}{L} \\ \dfrac{D_2}{C} & -\dfrac{1}{CR_{load}} \end{bmatrix} \begin{bmatrix} \bar{i}_L \\ \bar{v}_C \end{bmatrix} + \begin{bmatrix} \dfrac{V_{in}}{L} \\ 0 \end{bmatrix}
\tag{9.13}
$$

Referring instead to the equations representing the complete system model in terms of switching functions, and averaging the matrices and thus the functions, yields:

$$
\begin{bmatrix} \dfrac{d\bar{i}_L}{dt} \\ \dfrac{d\bar{v}_C}{dt} \end{bmatrix} = \begin{bmatrix} 0 & -\dfrac{D_2}{L} \\ \dfrac{D_2}{C} & -\dfrac{1}{CR_{load}} \end{bmatrix} \begin{bmatrix} \bar{i}_L \\ \bar{v}_C \end{bmatrix} + \begin{bmatrix} \dfrac{V_{in}}{L} \\ 0 \end{bmatrix}
\tag{9.14}
$$

which corresponds to what calculated above.

9.8.2   Discrete time models

Discrete time models for switching circuits are somehow a natural choice. The switching in fact occurs in general in a cyclic manner, and, as a consequence, a model that represents a sample per cycle of the variables of interest is particularly meaningful.

The functions that appear in the discrete time model cannot be immediately related to the continuous time model of the same circuit. There are cases, though, in which the connection between the two models is particularly significant.

The first such case comprises the piecewise linear circuits, i.e. a circuit switching between linear configurations. A second case is that of a very short sampling period, within which the model functions are not changing significantly. In this case, the solution model is achieved via approximation.

9.8.3   Generalized state-space models

These generalized models contain, besides the state variables, also auxiliary variables characterized by non-dynamic relations.

The auxiliary variables are mainly the instants of non-forced commutation, as for example that of diodes. Another relevant case is represented by switching that depends on the value of the state variables, such as the case of a feedback-controlled system.

The generalized model may include the control strategy, thus expressing in implicit or explicit way the dependency of the auxiliary control variables, e.g. the duty cycle, on state variables and inputs. A very relevant example is the hysteresis feedback control. In this case, the duty cycle cannot be considered an independent input, but rather it depends on the state variables.

9.8.4   Linearization of state-space models

The procedure to linearize state-space models to obtain small signal models is similar to the linearization of averaged models. The state variables are replaced with their expression in terms of nominal value and superimposed disturbance, and then the equations are expressed in terms of their first degree approximation. The model thus obtained has the same formulation as the state space model.

Let us consider first the continuous time domain model of the boost converter in figure. We report here the equations of the state-space model:

$$L\frac{d\bar{i}_L}{dt} = V_{in} - (1-d)\bar{v}_C$$
$$C\frac{d\bar{v}_C}{dt} = (1-d)\bar{i}_L - \frac{\bar{v}_C}{R_{load}}$$
(9.15)

The presence of products of state variables and control variables makes the model nonlinear from the control standpoint, making it impossible to apply the convenient methods of linear systems.

As described in the linearization process, we now express the state variables, inputs, and control variables as a sum of nominal value and disturbance, that is, the small signal:

$$\bar{i}_L = I_{L(nom)} + \tilde{i}_L$$
$$\bar{v}_C = V_{C(nom)} + \tilde{v}_C$$
$$V_{in} = V_{in(nom)} + \tilde{v}_{in}$$
$$d = D_{(nom)} + \tilde{d}$$
(9.16)

Substituting these variables in the state-space model yields:

$$L\frac{d\tilde{i}_L}{dt} = V_{in} + \tilde{v}_{in} - \left(1 - D_{(nom)} - \tilde{d}\right)\left(V_{C(nom)} + \tilde{v}_C\right)$$
$$C\frac{d\tilde{v}_C}{dt} = \left(1 - D_{(nom)} - \tilde{d}\right)\left(I_{L(nom)} + \tilde{i}_L\right) - \frac{V_{C(nom)} + \tilde{v}_C}{R_{load}}$$
(9.17)

Neglecting the terms which are products of two small signals, in steady state, the following holds:

$$V_{in(nom)} = \left(1 - D_{(nom)}\right)V_{C(nom)}$$
$$\frac{V_{C(nom)}}{R_{load}} = \left(1 - D_{(nom)}\right)I_{L(nom)}$$
(9.18)

And, finally, the equations take the following form:

$$L\frac{d\tilde{i}_L}{dt} = \tilde{v}_{in} + \tilde{d}V_C - \left(1 - D_{(nom)}\right)(\tilde{v}_C)$$
$$C\frac{d\tilde{v}_C}{dt} = -\tilde{d}I_{L(nom)} + \left(1 - D_{(nom)}\right)\tilde{i}_L - \frac{\tilde{v}_C}{R_{load}}$$
(9.19)

The objective is achieved, in fact the new averaged model is linear in the perturbations and retains the form of the state space model.

The small signal equivalent circuit of the boost converter can be represented from this model as in Figure 9.11.

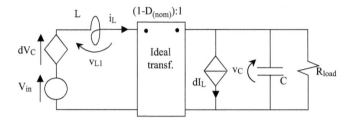

Figure 9.11   Small signal model of the boost converter

9.9   Implementing a switch

While the implementation of averaged models is basically immediate, for the switching behavior we considered three different options:

1. Ideal switch
2. Switching of parameter value
3. Switching of companion source

9.9.1   Ideal switch

The case of the ideal switch corresponds in the simulation implementation to a change of topology. If the number of switches is reasonably limited, this approach can be extremely efficient.

Both for the state equation case and the resistive companion, a set of models are stored in memory and an external logic defines which configuration is active at a given time. The price to pay is mostly in terms of memory: it is necessary to store 2^n options of models, if n is the number of switches.

While in the past, this option was considered unfeasible given the limitation of memory, it is now vice versa reasonable for a growing number of switches.

On the other hand, in a circuit-oriented simulation, this approach does not support the development of the switch as an independent model but the switch becomes integrated in the logic of the solver. This is, in general, a limitation for the flexibility of the platform and limits the options for the user in extending the libraries. On the other hand, it is extremely efficient from a computational point of view. Assuming that the switches are the only non-linear component in the system, using the ideal switch models transforms the circuit in a set of linear circuits. Each of them can be solved in a very efficient and fast way for example saving in memory the result of the LU factorization for each configuration.

9.9.2   Switching of parameter value

This option is particularly popular in topology-oriented solvers but it does not add value for a state space approach. The advantage of this approach is that the switch becomes a modular model as any other component of the network. There is no need of a special adaptation of the solver, and it does not introduce non-linearity rather time dependency.

At every time step, the component is linear but anytime there is a switching action the system matrix has to be recalculated. The computational burden is higher the matrix inversion may have to be recalculated several times at run-time. It is also possible to combine the previous concept with the one proposed here, memorizing the matrices with finite values for the parameters.

In fact, one of the main challenges of switch modeling is that the value of its conductance parameter theoretically oscillates between zero and infinity. In a nodal analysis approach, dealing with conductances, we have:

1. The switch is off, then the conductance is null.
2. The switch is on then the conductance is infinity.

In practice, solvers substitute the values with finite and not null values. Two options are typically used:

1. A purely resistive approach
2. A resistive-inductive approach

In the first case, the resistance of the switch changes between two finite values R_{min} and R_{max}. The value of R_{min} can be suitably selected to represent the conduction losses of the semiconductor, so it has a physical interpretation too. The main limitation of this approach is that the switching event significantly affects the eigenvalues of the circuit and this could affect the overall numerical stability of the solver. Numerical problems may emerge also from the use of values for the resistances that are extremely low or high.

To overcome these numerical issues, a possible solution is to switch jointly the value of an equivalent resistance and inductance so to preserve the time constant, i.e. the values are selected in such a way that:

$$\frac{L_{min}}{R_{min}} = \frac{L_{max}}{R_{max}} \tag{9.20}$$

9.9.3 Switching of companion source

A further evolution of the previous concept is given by the idea to consider the switch as a component behaving as a small inductance when it is closed and as a large capacitor when it is open. By properly selecting the value for C and L, it is possible to guarantee that the value of the equivalent conductance does not change for a given value of the time-step. In formula:

$$\frac{2C}{\Delta t} = \frac{\Delta t}{2L} \tag{9.21}$$

By using this process, the conductance matrix does not change in the presence of a switching event and, as result, the matrix inversion does not need to be updated. At every step, only the equivalent source in parallel needs to be properly updated depending on the switch condition.

This approach is extremely efficient from a software and computational point of view. In practice, it may anyhow create undesired oscillations in the waveforms.

For this reason, it can be convenient to add in parallel to the switch a "numerical snubber," i.e. the series combination of a resistance and a capacitance that smooth the transient. The following section presents the derivation of the companion model of the switches and the construction of the converter models.

9.10   Resistive companion model of converters

This modeling approach build on top of the companion modelling concept, see related chapter, to formulate a model of the switch in the Norton form that results in a constant admittance matrix of the circuit containing the switch. This implies that the admittance matrix does not change, irrespective of the status of the switch i.e. the topology of the circuit. As these models inherit the features of the companion models, they are discrete in nature and can be used to formulate the model of the circuit without following the integro-differential formulation and discretization steps. The model of a switching device is presented here, as developed in [2], formulated for the trapezoidal rule integration and in generalized form, to allow for other integration methods.

The starting point is the companion model of inductance L and capacitance C, which can be obtained by choosing $R=2L/h$ or $R=h/2C$ respectively, where h is the time step of the discretization, and inverting the direction of the current source. Exploiting the change of direction of the current source, it is then possible to model the behavior of the switch. The companion model of the idea switch looks like:

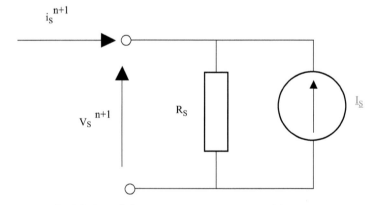

where $R_s=2L/h=h/2C$, and the current source $I=\pm I_s$ with

$$I = \begin{cases} +I_s \\ -I_s \end{cases} \quad I_s = \left[\left(\frac{2C}{h} \right) v_n + i_n \right]$$ (9.22)

The case $+I_s$ represents the situation in which the switch is off and the model behaves like a (small) capacitance, the case $-I_s$ represents the situation in which the switch is on and model behaves like a (small) inductance. It is evident now that the status of the switch can be changed by changing the direction of the current of the source of the companion model while the rest (the admittance matrix) stays the same.

An improved, stiffer, more stable version can be obtained by adding a parallel RC snubber to the previous ideal switch model. The combination of companion models of the ideal switch and the RC snubber results in the following:

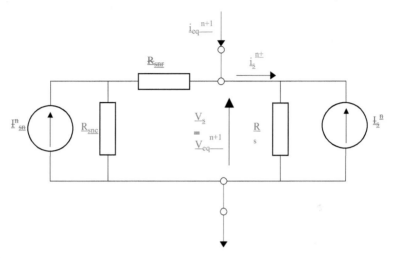

where,

$$v_s^{n+1} = v_{eq}^{n+1}$$

$$i_s^{n+1} = \frac{v_{eq}^{n+1}}{R_s} \pm I_s^n$$

Eventually:

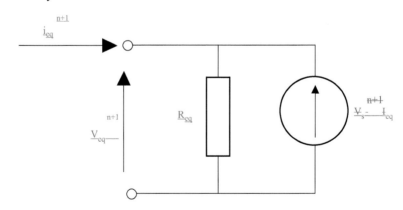

with:

$$R_{eq} = \frac{R_s(R_{snr} + R_{snc})}{R_s + R_{snr} + R_{snc}}$$

$$I_{eq}^n = I_s^n + \left(\frac{R_{snc}}{R_{snr} + R_{snc}}\right)I_s^n$$

This model can be generalized for other integration methods, as demonstrated in [2].

The use of the model of the switch with snubber is now demonstrated with the single-phase inverter in figure:

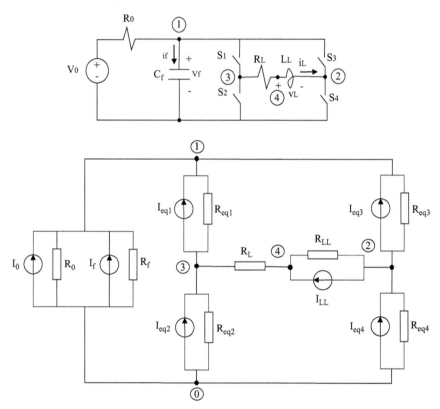

The switching circuit is modeled as in figure via:
Norton equivalent of the source, with:

$$I_0 = \frac{v_0}{R_0} \tag{9.23}$$

Companion model of the filter capacitance and load inductance have the following parameters:

$$R_f = \frac{h}{2C_f}, \qquad I_f = \frac{2C_f}{h}v_f + i_f$$

$$R_{LL} = \frac{2L}{h}, \qquad I_{LL} = \frac{h}{2L}v_L + i_L \tag{9.24}$$

The switches are modeled as the ideal switch with parallel snubber integrated with the trapezoidal rule as presented above. The load resistance is unchanged.

Eventually, the system model can be solved at every iteration with the nodal circuit equation $\mathbf{Y \cdot V = I}$ with the following:

$$\mathbf{Y} = \begin{vmatrix} R_0 + R_f + R_{eq1} + R_{eq3} & -R_{eq3} & R_{eq1} & 0 \\ -R_{eq3} & R_{eq3} + R_{eq4} + R_{LL} & 0 & R_{LL} \\ R_{eq1} & 0 & R_{eq1} + R_{eq2} + R_L & -R_L \\ 0 & R_{LL} & -R_L & R_L + R_{LL} \end{vmatrix}$$

$$\mathbf{V} = \begin{vmatrix} V_1 \\ V_2 \\ V_3 \\ V_4 \end{vmatrix} \tag{9.25}$$

$$\mathbf{I} = \begin{vmatrix} I_0 + I_f + I_{eq1} + I_{eq3} \\ I_{LL} - I_{eq3} + I_{eq4} \\ I_{eq2} - I_{eq1} \\ -I_{LL} \end{vmatrix}$$

In this chapter, various models of switching circuits are presented, and thus of power converters, suitable for use in device level control design and system analysis. Each model derivation and formulation demonstrate the challenges, sources of limitation of accuracy, and dynamics, in particular related to linearization and numerical issues. The user of the model should combine the purposes of the model with the characteristics discussed here to make a knowledgeable choice or model type and formulation.

Problems

1. Given a three phase DC/AC converter in which the DC is fed through a capacitor C and each phase of the AC side is fed through an inductance L, write the equation of a state space model using switching functions.
2. Given the topology of Problem 1, write the same model in terms of averaged model representation.
3. Connect the circuit of Problem 1 to a real voltage source on the DC side and to a three-phase resistance on the AC side and then write the possible configurations of the conductance matrix using the switching of parameter value approach.
4. Develop a resistive companion model of a switch based on the switching of companion sources and Gear2 integration.
5. Model a 2-switch resonant converter with switches based on the switching companion sources.

References

[1]   S. Y. R. Hui and S. Morrall, "Generalized associate discrete model for switching devices," *Proc. Inst. Elect. Eng.*, vol. 141, pt. A, pp. 57–64, 1994.
[2]   L. O. Chua and P. M. Lin, *Computer Aided Analysis of Electronic Circuits: Circuit Algorithms and Computational Techniques.* Englewood Cliffs, NJ: Prentice-Hall, 1975.

Chapter 10

Real-time and hardware-in-the-loop simulation

Christian Dufour[1] and Jean Belanger[1]

10.1 Introduction

Real-time simulators (RTS) have always been used to design and test actual control and protection equipment performance before their installation on actual power systems. Protection and control equipment were usually interfaced with analog benches emulating the real power systems in order to perform tests under realistic steady-state and faulty operating conditions. Actual control systems as well as analog benches were, by definition, operating in real-time meaning that control and protection systems under tests were exchanging signals at the same speed as when the tested equipment was installed on the actual power systems. The same technique was and is still applied in many other industries including automotive, aircraft and aerospace.

Over the last two decades, commercially available computers have become both increasingly powerful and very affordable. This, in turn, has led to the emergence of highly sophisticated simulation software tools enabling high-fidelity real-time simulation of dynamic systems, as well as fast electromagnetic transients (EMT) expected in electrical and power electronic systems. The consequence is that several expensive analog test benches and simulators have been replaced by more affordable, very flexible and accurate, fully digital RTSs. These RTS can then be connected in closed-loop with control and protection systems under design and test, as was the case with former analog simulators. This is called hardware-in-the-loop (HIL) simulation and test. Of course, the RTS must have the bandwidth and the very small latency required to simulate the high-frequency components of the phenomena that can affect the performance of the control and protection systems under test. The RTS must then have the ability to simulate all phenomena within a specified time step, from microseconds to milliseconds, depending on the application, as discussed in following sections. But in all cases, the RTS must maintain its real-time performance, i.e. all signals must be updated *exactly* at the specified time step. Otherwise, an overrun or a failure to update its outputs and inputs at the specified time step will cause unacceptable signal distortion affecting the performance of the equipment under test. Consequently, as seen from the terminals of systems under test, one

[1]OPAL-RT Technologies, Canada

should not be able to see the difference between the signals generated from the simulator and the real plan systems within a specified bandwidth.

Automatic code generation tools are used in conjunction with the current generation of a RTS for implementation in industrial controllers. These simulation software tools, including off-line and real-time simulation tools as well as automatic code generators, form the basis of the "Model-Based Design" paradigm; a control design methodology that is centered on the use of reference system models and simulation at each design step. In the model-based design (MBD) approach, described in greater detail in the next section, initial modeling and requirements, early controller prototypes, production code generation, production controller testing, and integration are all derived from reference models. The approach aims to accelerate the design cycle and reduce total design cost through the early detection of design flaws and other problems. In the automotive industry, in particular, the MBD method is a de-facto standard. MBD is also quickly gaining acceptance and being adopted by design engineers in a large number of industries.

Real-time controller prototypes, whose development is based on automatic code generation, are being used in many engineering fields and applications, such as aircraft flight control design & validation, industrial motor drive design, complex robotic controller design, and power grids. In several cases, the same controller code used for the prototyping phase is also used for the final hardware implementation.

The controller developed for these applications can be tested with scaled-down analog benches, as was common 15–20 years ago. However, these applications can now benefit from the use of fully digital RTSs in a number of ways.

• First, real-time simulation produces a set of requirements and specifications that can be used by all teams/subcontractors involved in a project.
• Second, it enables testing of prototypes and actual controllers at or beyond their normal operating limits without the risks involved in testing real devices connected to the actual systems being controlled, especially when high-power levels are present. It is easier and less risky to test fault responses on a simulated plant model.
• Finally, the simulation acceleration factor obtained by the use of compiled code (instead of the interpreted code used by most simulation tools), as well as by specialized parallel electrical solvers and hardware, enables the realization of rapid batch simulations. More tests can be done in less time.

Fast and real-time simulation can be used for statistical studies using the Monte-Carlo simulation method for power grid applications such as finding the statistical distribution of amplitude of overvoltage and over-currents caused by a series of random events. This analysis can be performed with prototypes or actual fast controller and protection equipment connected to the simulator (HIL).

10.2   Model-based design and real-time simulation

MBD is an engineering process that addresses problems associated with the design of complex systems and is based on the extensive use of models and simulation at

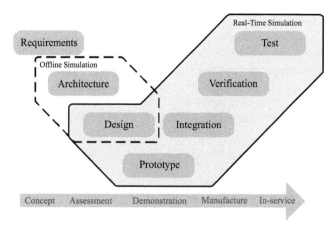

Figure 10.1 MBD workflow

each design phases. MBD is a methodology based on a workflow known as the "V" diagram, as illustrated in Figure 10.1. It allows multiple engineers involved in a design and modeling project to use models to communicate knowledge of the system under development, in an efficient and organized manner.

The left part of the V-cycle leads to the development of a production-type controller. The right part consists of steps to deploy this controller gradually until the final release.

MBD offers many advantages. By using models, a common design environment is available to every engineer involved in creating a system from beginning to end. Indeed, the use of a common set of tools facilitates communication and data exchange. Reusing older designs is also easier since the design environment can remain homogeneous through different projects. In addition to MBD, graphical modeling tools, such as the SimPowerSystems toolbox for Simulink® from The MathWorks®, simplify the design task by reducing the complexity of models through the use of a hierarchical approach.

Most commercial simulation tools provide an Automatic Code Generator that facilitates the transition from controller model to controller implementation. The added value of real-time simulation in MBD emerges from the use of an Automatic Code Generator (ACG). ACG translates graphical simulation models into "C" code. By using an ACG with a RTS, a hardware prototype of the controller – this is called Rapid Control Prototyping or RCP – can be implemented from a model with minimal effort. The prototype can then be used to accelerate integration and verification testing, something that cannot be done using offline simulation. The same holds true for HIL testing. By using an HIL test bench, test engineers become part of the design workflow earlier in the process, sometimes before an actual plant becomes available. For example, by using the HIL methodology, automotive test engineers can start early testing of a car or power system controller before a physical test bench is available. Combining RCP and HIL, while using the MBD approach, has many advantages:

- Design issues can be discovered earlier in the process, enabling required tradeoffs to be determined and applied, thereby reducing development costs.

- Development cycle duration is reduced due to parallelization in the workflow.
- Testing costs can be reduced in the mid- to long-term since HIL test setups often cost less than physical setups and the RTS can typically be used for multiple applications and projects.
- Test results are more repeatable since RTS dynamics do not change through time the way physical systems do.
- Can replace risky or expensive tests that use physical test benches.

10.3   General considerations about real-time simulation

10.3.1   *The constraint of real-time*

The most important aspect of real-time simulation technologies is that they need to make their computations fast enough to keep up with real world time. Moreover, this constraint must be respected at all times and, specifically in terms of solvers, at all time-steps. This means that the solver used to compute the equations of the phenomenon under study must be optimized with this in mind. Here, the expression "optimized solver" could be read as "simplified solver" in the sense that real-time solvers typically do not include basic DAE equation solver feature like iterations and error control. Recently, because of the increase of computational power available, real-time solvers sometimes include iteration nowadays.

Typically, the differential-algebraic equations (DAE) of the simulated system will be discretized and computed along a sequence of equal time-spaced points. Then, all these points must be computed and completed within the specified time step. If a time step is not completed in time, we get what is called an "overrun," meaning that the simulated process is late compared to real-time. Such overruns create distortion of the waveforms injected into the controller under test, which is not acceptable.

10.3.2   *Stiffness issues*

One other problem that occurs in simulating electric circuits is that their equations often exhibit stiffness. What is stiffness? Stiffness occurs when one simulates DAE with a spread in the eigenvalues or natural frequencies, let us say high, medium, and low, but with a special interest in the medium or "mid-range" bandwidth phenomena. Non-stiff solvers (e.g. explicit Runge–Kutta) will be unstable because of the higher frequency components of the DAE. In other words, to simulate these systems with a non-stiff solver, you will be forced to adapt the time step to the higher frequency component and the simulation will become extremely long for the frequency of interest. But a certain category of solvers called "stiff solvers" are able to "cut through" the high-frequency components and be less influenced by frequency components higher than sampling frequency. Mathematically speaking, the solver stiffness property refers to A-stability or L-stability.

For power systems switching transient studies, people are typically interested in components below 2 kHz. But the equations may have eigenvalues in the MHz range that cannot easily be eliminated. For example, modeling switches with a small- and a high-resistor value to simulate the on- and off-state have a tendency to create stiff systems. This is the reason that most power system simulator solvers are based on the A-stable order-2 trapezoidal rule of integration, like in RTDS [1] or Hypersim [2]. However, in most cases, users must still use snubber resistors and capacitors across the switches to improve numerical stability. Newer discretization methods as emerged like the order-5 L-stable discretization rules of ARTEMiS-SSN [3]. Because of its higher order, ARTEMiS order-5 can be more accurate than order-2 solver for the same time step value. Such features enable the use of larger time step values, which facilitate real-time simulation of difficult cases.

10.3.3 Simulator bandwidth considerations

Another important aspect of a RTS is the bandwidth. This is to say that the definition of the simulation (the selected time step) must be compatible with the speed or bandwidth of the phenomena that we simulate. For example, simulating electric circuits requires a much smaller time step than mechanical ones because the former is much faster by nature. This is one reason why electric systems are difficult to simulate: their bandwidth is high, imposing smaller time steps and thus more powerful computers to simulate them.

Because the simulation is a sampled process, the appropriate bandwidth of a simulator refers to the Shannon theorem, which requires a minimum of two samples within the period of the largest frequency of interest. But in practice, the simulation sampling frequency (inverse of the time step) will be selected between 5 and 10 times the frequency of the phenomenon under study.

It is a common practice with EMT simulators to use a simulation time-step of 30–50 μs to provide acceptable results for transients up to 2 kHz. Because greater precision can be achieved with smaller time-steps, simulation of EMT phenomena with frequency content up to 10 kHz typically requires a simulation time-step of approximately 10 μs.

Accurately simulating fast-switching power electronic devices requires the use of very small time-steps to solve system equations. Offline simulation is widely used, but is time consuming if no precision compromise is made on models (i.e. the use of average models). Power electronic converters with a higher PWM carrier frequency in the range of 10 kHz, such as those used in low-power converters, require time-steps of less than 250 ns without interpolation [REF FPGA], or 10 μs with an interpolation technique [REF MELCO]. AC circuits with higher resonance frequency and very short lines, as expected in low-voltage distribution circuits and electric rail power feeding systems, may require time-steps below 20 μs. Tests that use practical system configurations and parameters are necessary to determine minimum time-step size and computing power required to achieve the desired time-step. Onboard drive protection systems definitively require at least 1 MHz bandwidth (and FPGA technology); these include for example ultra-fast systems like DC-bud short-protection.

10.3.4 Achieving very low latency for HIL application

One of the main applications of RTS is to develop and test protection and control systems by interconnecting the equipment in closed-loop with the simulator. This is called hardware-in-the-loop (HIL) or controller-in-the-loop (CIL). In such application, the total system must react like the real systems, implying that the simulator delays are small compared to the delays and time constant of the actual controller, which is usually the case for typical thyristor-based power electronic controllers when the simulator time step values are 25–50 µs.

However, the use of fast voltage source converters (VSCs) with PWM frequency larger than 10 kHz as well as multi-modular converters (MMC) with a very large number of levels my required time step values below 1–2 µs to achieve a total latency below 2 or 4 µs. The latency is defined as the time elapsed between the insulated gate bipolar transistor (IGBT) firing pulses sent by the controller under tests and the reception, by the controller of voltage and current signals sent back by the simulator. Reaching such a low latency requires the use of field programmable gate array (FPGA) chips that are available in most modern simulators.

10.3.5 Effective parallel processing for fast EMT simulation

Conventional off-line simulation tools such as EMTP, EMTP-RV, ATP, and PSCAD typically use only one processor to compute the total system, which means that the time to compute one time step can exceed the wall-clock time when the size of the simulated system increases. However, real-time constraints also require the use of highly optimized solvers, taking advantage of parallel processing to reach the specified time step even if the complexity and size of the simulated system increases.

An important improvement in simulation speed was achieved in the 1990s by using the transmission lines modeled with distributed parameters to make the admittance matrix block-diagonal, creating subsystems that can be solved independently from each other. This property allows dividing the network into subsystems with smaller admittance matrices, which can be solved in much shorter time. The resulting simulation model structure is illustrated in Figure 10.3 where the line equations serve as links between the decoupled set of equations.

However, the effective implementation of this technique implies that the processing time for data exchange between processors is much smaller than the time used to simulate each subsystem. Fifteen years ago, only high-end commercial or custom-made very expensive super computers were able to implement such parallel simulation effectively with time steps as low as 50 µs. The main impediment was the difficulty in implementing a low latency inter-processor communication fabric capable of transferring data with a delay below 50 µs without overloading each processor. In the early 1990s, organizations such IREQ (Hydro-Quebec's R&D center), EDF (ARENE simulator [4]), and Manitoba HVDC R&D centers (and RTDS Inc.) developed their own custom made super computers to implement real-time digital parallel simulators. Such super computers were expensive to develop and to maintain.

By contrast, PC and laptop computers usually found in every office today are equipped with processors that have up to 8 cores. Standard PC servers can be equipped with up to 32 processor cores tightly interconnected by a fast shared memory. They are able to run several threads in parallel with fast inter-processor communication, provided the software uses these features.

Traditional EMT off-line simulation software, such as EMTP and PSCAD, are being modified by their respective development teams to take advantage of parallel processing. The EMT simulation software packages, Hypersim, eMEGAsim, and RTDS have been specifically developed, from the beginning, to take advantage of parallel and distributed processing, with great improvements in simulation speed. To perform this, Hypersim and RTDS partition the network at the connection points to transmission lines, creating subsystems, as illustrated in Figure 10.2.

Hypersim and eMEGAsim generate C-code for each subsystem and compile them individually. Linkage is performed, depending on the commercial hardware used for the simulation, so that different executable modules are generated to be assigned to different processor cores. In this way, the simulation speed is increased according to the number of assigned processor cores.

Over the last 20 years, RTDS Inc. has developed more and more powerful custom computer boards optimized for real-time simulation power grids. At the same time, the same Hypersim EMT parallel code has been ported from custom super computer, to SGI super computer and to standard multi-core PCs.

Figure 10.3 illustrates an example of a small but powerful 4-core simulator that can simulate a power grid with about 220 electrical nodes (single-phase) at 50 µs

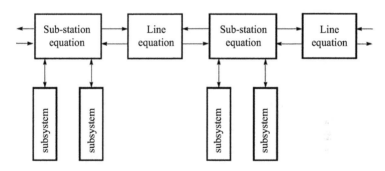

Figure 10.2   Program execution structure for EMT parallel simulation

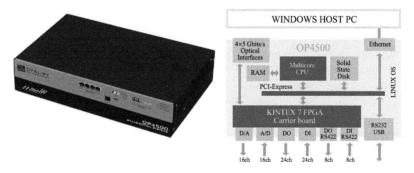

Figure 10.3   A multiprocessor desktop computer for fast simulation of complex systems

using a standard and low-cost INTEL i7 4-core processor that can be found in typical PCs. This small desk-top simulator is also equipped with a powerful KINTEX-7 FPGA to simulate fast power electronic converter systems and controllers with time step below 1 μs [5]. This setup can run a small model in real-time, for testing purposes, with HIL.

Although Hypersim is adequate for regular PC's, an interesting alternative for large, complex simulations is using server-graded multiprocessor computers, which are now available at costs comparable to normal workstation prices. In this case, a 32-core INTEL based server can be used to simulate grids with more than 2,000 electrical nodes with hundreds of generators, transformers and loads, and several high-voltage direct current (HVDC) transmission systems with a time step of 50 μs. This capability is very useful for integrating test of power grids equipped with several HVDC, flexible AC transmission system (FACT), and distributed generation.

For very large grids, Hypersim can also take advantage of large SGI super computer with hundreds of processors cores. The capability to run the same simulation software on laptops or on commercial servers, super computers, and on the cloud is certainly an advantage for large utilities, R&D centers, and universities.

It must also be noted that using parallel processing for EMT simulation is not only essential for real-time HIL simulation but also to accelerate simulation studies performed in off-line mode. A faster simulation speed improves user interaction and enables analysis of more contingencies in less time. For such applications, Hypersim can take advantage of commercial or in-house cloud computing.

Of course, using hundreds of processor cores to simulate very large grids requires the use of efficient and automatic processor allocation software to facilitate to use of such powerful parallel computer. Hypersim provides this essential feature with the capability to calculate all initial conditions.

10.3.6 FPGA-based multi-rate simulators

The simulation of complex VSCs used in transmission and distribution systems such a modular MMC, as well as fast converters found in distributed generation systems such as PVs, wind farms, and micro grids often require simulation time steps below 1 μs. Such a small time step cannot be achieved by today's standard computer. Specialized processor technologies such as FPGA processor must be used simulation time step as low as 200 ns and to interface simulation results with external equipment using fast I/O converters. An example of such simulator architecture is presented in Figure 10.3 illustrating the OPAL-RT OP4500 simulator. In this case, the slower subsystems are simulated with a time step of 10–50 μs on standard INTEL i7 processors while the fast power electronic systems are simulated with a time step below 1 μs. A general nodal solver, called eHS (electrical **H**ardware **S**olver), has been implemented directly on FPGA chips to facilitate the simulation of complex power electronic systems [5,6]. Users can simply draw the circuit diagram and the FPGA will automatically simulate it without any need to learn complex FPGA programming.

A powerful multi-FPGA simulator has also been delivered for the real-time simulation of HVDC MMC power grids using a total of 7,500 MMC cell models and 5 converter terminals [7]. Such FPGA simulators are interfaced with eMEGAsim and Hypersim multi-core simulators to provide the capability to simulate very large power grids integrating fast power electronic systems for transmission, distribution, and micro-grid systems.

10.3.7   Advanced parallel solvers without artificial delays or stublines: application to active distribution networks

The parallel solver implementation described above is based on the availability of transmission lines or cables to separate the system into small independent subsystems, which can be simulated within the specified time step using only one processor. If the computational time for one of these subsystems becomes too large because it has too many nodes, then the common practice is to add artificial delays to enable parallel processing to reduce the computational time.

An artificial delay is usually implemented with a stub line, which is a line with a line length adjusted to obtain a propagation time of one time step. Large capacitors and inductors can also be used to split a large system in several smaller subsystems to take advantage of parallel processing. However, adding artificial delays also adds parasitic series inductors and parasitic shunt capacitors. Users must then compare results obtained without the addition of artificial delays using off-line simulation with results obtained using the artificial delays to reach the specified time step. These parasitic L and C can be tolerated in many applications depending on the values of these elements with respect to the other circuit impedances. This technique is used successfully by RTDS, eMEGAsim, and Hypersim users for many cases.

However, in some cases, like large distribution circuits simulated with several PI circuits, the addition of artificial delays may be problematic as parasitic L and C will be large compared to actual component values and therefore the standard decoupling methods cannot be used in distribution systems. Simulating large converter systems interconnected with a large number of AC and DC switchable filters may also be problematic to reach time step values below 25–50 μs even with the fastest available processor. Steady-state and transient values may be affected by the addition of artificial delays. Consequently, the use of circuit solvers that can simulate large circuits in parallel without adding any parasitic L and C can be very useful to increase simulation speed and accuracy. This is the key feature of the state-space-nodal (SSN) algorithm described in detail in [3,8].

SSN is a nodal admittance-based solver that allows minimizing the number of nodes and thus the size of the nodal admittance matrix. This, in return, can help increase the speed of simulation because the LU solution of the admittance equation requires a number of operations that is proportional to the cube (N^3) of the size N of the nodal matrix. It does so by letting users choose the node locations and the corresponding groups of elements.

As a simple example, we show the 3-phase circuit of Figure 10.4. In this circuit, the classic method to build the nodal admittance matrix, using standard

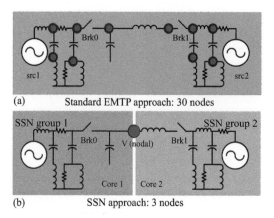

(a)   Standard EMTP approach: 30 nodes

(b)   SSN approach: 3 nodes

Figure 10.4   Comparison of node number for standard nodal admittance method and SSN

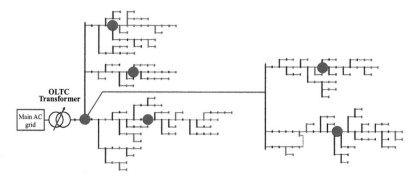

Figure 10.5   Large distribution network with more than 650 "EMTP-type" nodes

RLC branches, would result in 30 nodes. By comparison, the SSN method can result in only three nodes. The reduction of the size of the nodal admittance matrix comes with an increased complexity in the resulting SSN group equations (an SSN group can be viewed as a multi-terminal generalization of a basic EMTP branch).

When the SSN group equations are large enough, they can also be computed in parallel on different processor cores, without delays, thus accelerating the simulation even more. The SSN algorithm is more complex to compute than the simple nodal technique but actual results show a performance gain when using up to four processors, which can be very useful for the simulation of difficult cases where too many R, L, and C components are tightly interconnected without lines or cables. This is depicted in Figure 10.6 (in the next section about the iterative MOV feature) where we can observe that the SSN group history source calculation and update stages are threaded on different cores without delays. For the sake of clarity, it must be understood that a similar delay-free parallelization could, in theory, be done

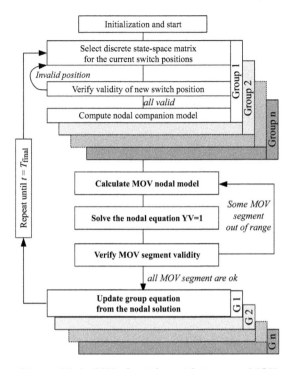

Figure 10.6   SSN algorithm with iterative MOV

with the standard EMTP approach. But in practice, if the branch size is too small and their number too high, the parallelization gains would be nullified by inter-processor communication overhead and other phenomena like CPU cache trashing.

Using the SSN approach and the principles shown in Figure 10.4, it is possible to simulate in real-time a radial distribution system with more than 700 nodes (with 980 L–C states from short pi-lines) at a time step under 65 μs, *without algorithmic delay or stublines*, using 4 cores of a standard Xeon multi-core server running RT-LAB. This performance is possible because the SSN reduces the network into a system with only 6 nodes and 6 multi-terminal branches (i.e. SSN groups) and uses threaded process to compute the SSN group equation in parallel, without any delay. A similar distribution system, depicted in Figure 10.5, with an OLTC at the feeder point and more than 650 equivalent EMTP nodes can also be simulated in real-time with the SSN solver. Using SSN and an 18 SSN nodes separation of the network (the 6 red dots in the figure), this model can be simulated in real-time at a time-step of 55 μs on a 3.3 GHz i7 Intel PC, *without any delays or stublines*. In [9], this network was simulated without the SSN solver and some delays were added to parallelize the simulation.

10.3.8   The need for iterations in real-time

When simulating in real-time highly non-linear devices such as surge arresters and metal oxide varistors (MOV), it is often necessary to use iterations in the standard nodal admittance solver. Hypersim [10] and ARTEMiS [11] implement this type of iterative algorithms. The previous reference clearly shows realistic cases in which iterations are required to obtain accurate results.

Making iterations in a RTS can be done efficiently because MOV only modifies the YV=I equation of the solver, without history source computations. This is depicted in Figure 10.6.

Similar iterative principles are currently being applied to switches and also, for Hypersim only, saturable transformers.

In a recent paper [12], this iteration feature was demonstrated to be required for accurate simulation of MOV-protection MMC converters acting as STATCOMs.

10.4   Phasor-mode real-time simulation

Real-time simulation for EMT studies in power system has been well established. As discussed in previous sections, the fundaments of most power grid real-time simulations are based on mathematics of EMT type analysis. This includes the hardware/power/model-in-the-loop simulations. However, power system experts long ago realized the fact that besides the real-time EMT domain simulation, they also need tools for off-line real-time phasor domain simulation. These types of tools can be used to test the functionality of hardware such as global control and special protection devices in large-scale power systems. Furthermore, it can be utilized for training purposes in academic laboratories, or as the operator training tool in energy management centers. Some organizations are contemplating the use of faster-than-real-time phasor-type simulators for on-line prediction of system instability and implementing corrective actions to prevent system collapse or minimizing damages.

The ePHASORsim from OPAL-RT Technologies Inc. is a simulation environment for both transmission and distribution power systems for balanced and unbalanced systems [13]. The mathematic background of ePHASORsim is based on a set of DAEs as follows:

$$\dot{x}(t) = f(x, V) \tag{10.1}$$

$$YV = I(x, V) \tag{10.2}$$

$$x(t_0) = x_0 \tag{10.3}$$

where x is the vector of state variables, V and I are the vector of bus voltages and currents, Y is the nodal admittance matrix of the network, and x_0 is the initial values of state variables. The solver core is built as a MATLAB®/SIMULINK® S-function and its library of models are coded in C++ that can also be used in standalone mode. The built-in library includes major and most common components that are used for this type of simulation in power systems, such as synchronous machine and its controllers, different type of loads, and transformer with on-load tap changers. An interface with Modelica FMI enables users to implement their own models. User models can also be implemented in C++ or with SIMULINK.

As of 2014, ePHASORsim could simulate systems with up to 10,000 nodes with a time step of 10 ms with only one Intel processor core. The systems can include about 3,000 synchronous machines (6 states), 5,000 controllers and 20,000 other components. Faults, line switching and OLTC can also be simulated. This number has been increased 10-fold as of late 2017 with ePHASORsim being able to simulate approximately 100,000 nodes at 10 ms.

A similar research is being done by PNNL in the USA [14] and, in Europe, by PEGASE for example [15], to develop real-time and faster than real-time parallel transient stability simulators.

Real-time phasor-type simulation can be used efficiently to test voltage control in large distribution system for example.

10.5 Modern RTS requirements

Modern RTSs are typically built around multi-core multi-processor (CPU) PCs with extensive I/O capabilities based on FPGA technology. Some simulators are using custom-made computer boards (see for example Figure 10.7).

A central processing unit (CPU) is a highly serialized arithmetic processing unit with a very flexible code flow architecture that enables it to implement very complex algorithms such as advanced ODE-solvers. In modern PCs, it is

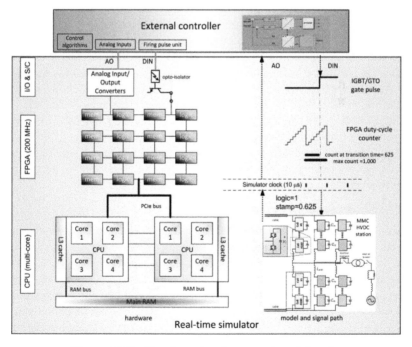

Figure 10.7   Simulator hardware and model paths

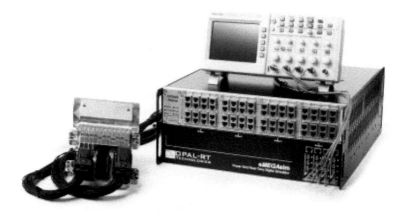

Figure 10.8   OP5600 RTS (right) with a controller under test (left)

commonly made with several computing cores and interfaced with other I/O components through a bus structure, like PCIe in PCs.

An FPGA is a massively parallel structure of basic logic and memory elements that can be assembled into computing units from a user specification. Typical FPGA chips are mounted on electronic boards that, in addition to RAM memory and SSD hard disk, provide a direct and rapid interface with I/O points. It can also implement models and solver of moderate complexity, especially for power electronics [5] as well as fast control systems and signal processing.

In Figure 10.8, we see the simulator hardware composed of CPUs and FPGA, along with the model path for an MMC (Multi Modular Voltage-Source Converter) circuits. On the figure, we see that the I/O signals driving the IGBTs of the MMC cells are first captured by some timing on the FPGA card. This allows keeping a precise timing of the IGBT gate pulse. This timing, a time stamp, is then sent to the complete MMC model on the CPUs, where the time stamp is used to correctly compute the model voltages and currents.

In other cases, the MMC models can be simulated directly in the FPGA chips [7] using a very small time step value below 1 μs to increase overall accuracy and to decrease the total latency.

10.5.1 Simulator I/O requirements

I/O requirements are increasing for real-time simulation. Today's power electronics systems are getting more complex as news topologies, like the MMC, gain acceptance. In order to perform HIL simulation of power electronic devices, RTSs must have various I/O types such as time-stamped digital Input – capable of sampling switching device gate signals with a resolution better than 100 ns. These types of inputs are sometimes referred as "PWM inputs, Analog Output" (1 μs sampling time) – to emulate the device current, voltage, and sensors as viewed by the controller under test – and Digital Input and Output to for routing various signals between the controller under test and the real-time emulated power electronic devices.

The FPGA also provides the capacity to easily code and emulate various sensors, like motor resolvers, quadrature encoders, RVDT, and including fault sensors.

Modern power electronic devices and controllers can also communicate through high-level communication links, such as Ethernet-based IEC-61850, DNP3 protocols for power system relays and substations, C37.118, the IEEE standard for using synchro-phasors in power systems, CAN protocol for automotive or TCN protocol for trains. The simulator must be able to physically interface with such protocols with an appropriate driver. For this purpose, on PC-based RTSs, the simulator can be equipped with a PCI/PCIe interface board and/or a simple Ethernet board for IEC-61850. Alternately, the user can also program the FPGA board to interface to the desired protocol.

The simulator must also provide proper signal conditioning for all I/Os such as filtering and isolation. Additionally, the simulator should provide easy access to I/O interfaces between the simulator and the device under test. Typical RTSs provide such probing points on the front panel of the simulator, like the OPAL-RT OP5600 simulator (Figure 10.9), or with a more complete patch panel. On the figure, we see the probing connection points on the front of the simulator. The controller under test is interfaced with the simulator through the I/O interface on the back of the simulator.

The ability to provide a very high number of I/O points is also very important in some applications. The MMC is a good example of device with a very large number of I/O.

As an example, a dual three-phase 100 cells/arms MMC simulator was recently commissioned by OPAL-RT with a total of 2,788 I/O points with 3 digital signals per cells (1,800), 1 analog output per capacitor (600) and various other voltages/ currents sensors and well as breakers signals. The time to manage all these I/O channels and to compute the models was 25 microseconds using parallel processors and several FPGA chips.

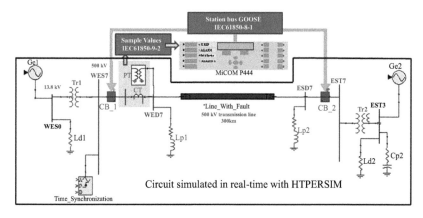

Figure 10.9 Typical protection relay testing setup with IEC61850 and HYPERSIM

A large MMC system, with up to 5 converters with 2,400 cells each (120,000 cells in total) to simulate a DC grids, was also delivered to China Institutes using 10 FPGA systems – OP7020 Virtex-7 FPGA racks that basically extend the number of I/O for the OP5600 simulator – to simulate all MMC in less than 500 ns and to interface thousands of signal to customer controllers using optical fibers.

10.6   Rapid control prototyping and HIL testing

Various applications of real-time simulation are detailed in the next sections. These applications can be broken into three important categories:

- Rapid control prototyping (RCP)
- HIL testing or CIL
- Software-in-the-loop (SIL)
- Power-hardware-in-the-loop (PHIL)
- Rapid simulation (RS).

Consider a controlled process that is composed of a plant with a controller acting upon it. In RCP applications, an engineer will use a RTS to quickly implement a controller and connect it to the real plant or to a simulated plant. HIL (or CIL) acts in an opposite manner; its main purpose is to test actual controllers connected to a simulated plant. SIL can be done when controller object code can be embedded in the simulator to analyze the global system performance and to perform tests prior the use of actual controller hardware (HIL or CIL).

PHIL consists of using real power component in the loop with the simulator. An example is a high-power motor emulator used to test actual inverter systems. The motor emulator uses a RTS controlling a power amplifier to output the simulated motor current with the same amplitude as the actual motors. PHIL is also used for battery emulator and micro-grid laboratories [16–19].

RS take advantage of parallel processing to accelerate simulation in massive batch run tests, such as aircraft parameter identification using aircraft flight data or in Monte-Carlo simulation used in power system analysis. RS is very useful for simulating large MMC HVDC grids using detail converter models to reduce simulation time by several orders of magnitude depending on system complexity and size.

10.7   Power grids real-time simulation applications

10.7.1   Statistical protection system study

Protection and insulation coordination techniques make use of statistical (Monte-Carlo) studies to deal with inherent random events, such as the electrical angle at which a breaker closes, or the point-on-wave at which a fault is applied. For protection coordination studies, multiple fault scenarios are required to determine appropriate protective relay settings and correct equipment sizing. By testing multiple fault occurrences, the measured quantities are identified, recorded, and stored in a database for later retrieval, analysis, and study.

Protection relay development and testing is one of the most traditional uses of RTSs. Actual protection relay equipment are interfaced to the simulator using voltage and current amplifiers in such a way that the relays received the same signals as if they were connected to actual power systems. Protection relay breaker tripping signals are connected to the simulated breaker through logical inputs. Faults can then be applied to the simulated network to test relay performance i.e. its operating speed and accuracy as well as to test the relay security i.e. to determine the number of erroneous operations.

Figure 10.10 presents a typical setup for testing modern relays interfaced with the IEC 61850 Ethernet protocol replacing hard wired connections.

Data measured by the PT and CT are sent by the simulator using Ethernet protocol IEC 61850-9-2 Sampled Values to the MiCOM P444 relay. The relay backup overcurrent function detects the fault and the relay sends an IEC 61850-8-1 GOOSE message back to the simulator in order to control the two circuit breakers at both ends of the transmission line.

Many automated tests must be executed to evaluate the relay performance as function of the fault impedance, duration, and incidence angle.

Figure 10.10 presents such statistical results, executed with TestView, the HYPERSIM test automation software, by modifying all these parameters so a fault is provoked at 50 km from the west bus on the 500 kV transmission line. In this test, the incidence angle varied between 0 and 359 degrees and 332 tests were run in order to characterize the reaction time of the relay. The average value was 21.25 ms and the standard deviation was 1.08 ms. Several tests must also be performed to test the capability of the relay to NOT operate for transient voltages and current values outside it specified operating range. Of course more complex tests can be performed using several relays installed on several lines and transformers to evaluate the risk of cascaded relay miss operations caused by the effect of previous relay operation. Such tests are often performed to test new relay design or new network topologies. The integration of new HVDC and FACT technologies as well as distributed generations and micro-grids may require more sophisticated tests.

In several cases, transients overvoltage and overcurrent waveforms can be recorded and played back later to test the relays in open loop mode. This testing method is very popular when it is not necessary to evaluate the effect of relay

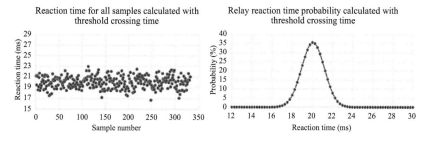

Figure 10.10   Protection relay testing results showing the probabilistic reaction time

operation on the performance of other relays. Waveform recording can be performed by standard off-line simulation software, such as EMTP-RV and PSCAD. RTSs are also used in faster-than-real-time mode, when the grid is very large, to reduce the time to record thousands of waveforms.

10.7.2   Monte Carlo tests for power grid switching surge system studies

The Monte Carlo technique has been used in the past for practical studies, such as the evaluation of the probable overvoltage at a substation [20], using traditional offline simulation software such as ATP, PSCAD or EMTP-RV as well as RTSs.

The large power system model depicted in Figure 10.11 illustrates a network with a very large number of busses and sort lines. It was first built and tested in the EMTP-RV environment and then converted to the SimPowerSystems/Simulink environment as a distributed model, ready to use with the RT-LAB RTS to perform statistical tests in less time but also to analyze the impact of SVC and futures HVDC systems using actual control equipment.

The 60 Hz, 138/230 kV HVAC power system model is an 86-bus electrical network. Its large number of transmission lines supply power to a total of 23 loads and 9 ideal voltage sources with lumped equivalent impedance represent the generators. Full machine dynamics can easily be added, as these models are available in SimPowerSystems.

Figure 10.11 also describes the simulated power system along with CPU task separation for the 8-core target used in this test.

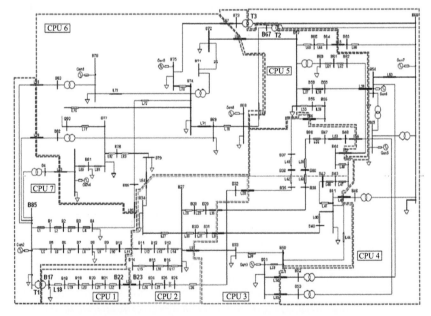

Figure 10.11   Schematic diagram of the network model – courtesy of Entergy

Two decade ago, simulation of such a model would not have been possible. A model of such complexity would have required a supercomputer composed of proprietary hardware. Today, this model can be simulated in real-time at a time step of 50 μs on the RT-LAB real-time simulation platform. This particular 86-bus model can now be run on four (4) 3-GHz cores.

The Monte Carlo test is designed using the Python script language, and using the RT-LAB API to control simulation runs, data acquisition, and post-treatment. As can be observed in Figure 10.12(A), at the point where two different fault durations are applied (1 cycle versus 3 cycles), the overvoltage measured during certain contingencies is dependent on fault timing. This comparison shows the necessity of using statistical methods to calculate the maximum overvoltage to achieve an efficient design of the power system under study. It is also important to understand that a significant number of tests may be needed, depending on the study. The total number of tests needs to be large enough to obtain acceptable precision on the statistic distribution. The criteria used to determine the number of tests can be decided upon by calculating the evolution of the average value (V50%) and/or, for instance, the evolution of the maximum overvoltage having a 2% probability of occurring (V2%). The evolution of these two quantities, according to the number of tests performed, is illustrated in Figure 10.12(D).

The above study can be performed in off-line mode using conventional software or with parallel software such as Hypersim or eMEGAsim to execute more tests in less time. However, in several cases, actual protection or controller hardware must be interconnected with the simulator to analyze the effect of fast control and protection

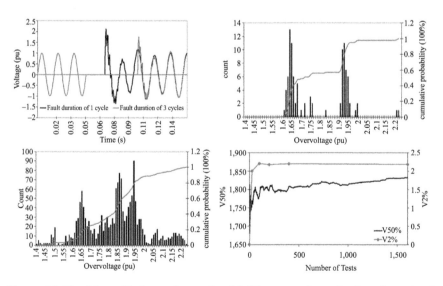

Figure 10.12   Monte Carlo study results. (A) Phase-a voltage for 1-cycle and a 3-cycles fault duration (Tstart = 50 ms). (B) and (C) Overvoltage statistic distribution for 100 and 1,600 tests. (D) Mean value and standard deviation evolution with respect to number of samples

actions on system overvoltage, overcurrent and arrester energy. Such real-time statistical tests also enable testing of control and protection system performance under
realistic transient conditions and discovering unwanted random controller operations.

10.7.3   Multi-level modular converter in HVDC applications

The inclusion of HVDC and FACTS devices in electric power grids is expanding
rapidly. Also relatively recently, the use of VSCs based on IGBTs is becoming more
attractive mainly due to their cost and higher performance. The recent Modular
MMC topology, based on half-bridge modules connected in series [21], offers significant benefits compared to previous VSC technologies, such as two-three level and
neutral-point diode-clamped (NPC) topologies. By using a significant number of
levels per phase in the MMC, the filter requirements can be eliminated. Moreover,
switching frequency and transient peak voltages on IGBT devices are lower in the
MMC, which reduces converter losses. An HVDC systems made by MMCs is shown
in Figure 10.13. Scalability to higher voltages is easily achieved and reliability is
improved by increasing the number of sub-modules (SMs) per phase.

The excessive number of power switches in the MMC creates significant
computational difficulties in EMT-type simulation tools. The numerous and non-
linear devices in the converter require an iterative process to solve the global matrix
which significantly increases the computational burden. Thus, in real-time simulations, modeling a highly accurate switching device is out of reach with the current
computational technology and some form of simplification is required to accom-
plish network integration and HIL studies.

Real-time simulation of these large MMC systems enables users to develop
and test actual controller equipment by interfacing them to the simulator in closed-
loop. Such an HIL set-up makes it possible to test the control system performance
under several operating conditions before its installation in the real power grid.
RTSs with the ability to simulate MMC systems and HVDC grids are therefore
essential and used by MMC manufacturers, R&D centers, and utilities.

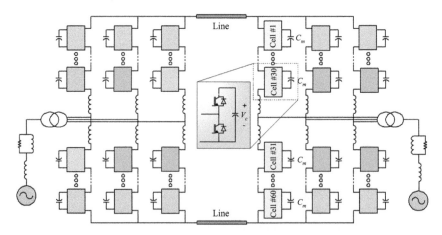

Figure 10.13   An MMC-based HVDC link

In [22], full real-time digital simulation of a static modular MMC HVDC link interconnecting two AC networks is discussed. The converter has 60 cells per arm; each cell has two power switches with antiparallel diodes and one capacitor. The simulated model can be used to study the natural rectifying mode, which is very important in the energizing process of the converter, whether a ramping voltage or a charging resistance is used. Current injection with delay is the method used in this approach. This facilitates the use of HIL in realistic sized MMC system that can have more than a thousand cells.

In [23], the MMC is simulated using the more traditional nodal admittance approach of SSN. The advantage of this approach is to avoid injection delays that can destabilize the simulation. The difficulty in using this approach is to effectively include the I/O signals with small latency when the cell number increases.

10.7.4 High-end super-large power grid simulations

Because of the importance of grid reliability, utilities have very stringent testing requirements and many of them developed their own RTS for these purposes. Hypersim, the Hydro-Quebec Research Institute (IREQ) RTS, is a good example of these high-end simulators, which was developed and used over the last 20 years.

As an example, the complete power network of the Province of Quebec, depicted in Figure 10.14, including 25 DFIG-based wind power plants, was simulated in real-time on the Hypersim RTS [2]. The network contained the following

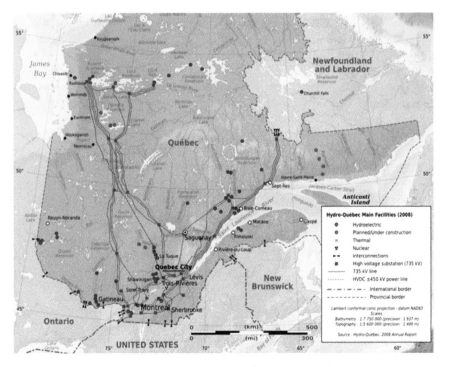

Figure 10.14   Quebec grid simulated in real-time by Hypersim

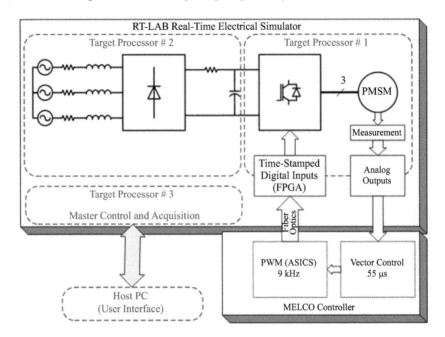

Figure 10.15   Real-time simulation of a PMSM drive

elements: 643 three-phase buses, 34 hydroelectric generators (turbine, AVR, stabilizer), 1 steam turbine generator, 25 Wind Power Plants with DFIG generators, 7 static VAR compensators, 6 synchronous condensers, 167 three-phase lines, on multi-terminal DC-link, and 150 3-phase transformer with saturation modeling.

Hypersim uses today (2018) 32 processors in an SGI super-computer to perform the real-time simulation of this network at a time-step of 50 μs. Larger systems are simulated by China State Grid using Hypersim [24].

One key aspect of this type of "super-large" EMT simulations is that the Hypersim simulator comes with automatic task allocation. The Hypersim graphical interface is also designed to cope with super-large grid diagrams including, single-line schematics as only one example. Continuously developed and improved internally at IREQ since the mid-1990s, Hypersim is now marketed by OPAL-RT Technologies since 2012.

10.8   Motor drive and FPGA-based real-time simulation applications

10.8.1   Industrial motor drive design and testing using CPU models

Mitsubishi Electric Co. has used real-time simulation technologies to design motor drives for machine tool applications. The challenge faced on this project was that the motor itself, a permanent magnet synchronous motor, and its related controllers were being designed simultaneously. Therefore, a physical motor was not available for controller tests. The solution was to use a virtual motor simulated in real-time during the controller testing phase [25].

The setup, depicted in Figure 10.15, consists of two main parts: the controller and the motor drive circuit. The controller includes a control module and PWM generator board. The vector control runs at 55 μs and the PWM carrier frequency can be varied up to 9 kHz. The motor drive circuit was implemented in Simulink and simulated in real-time on an RT-LAB electrical simulator. The motor drive was simulated more than 10 years ago by two Pentium 4 target processors, each operating at 2.8 GHz: one for the AC–DC part, and one for the DC–AC part, including the motor. A third Pentium processor was used for master control of the simulator, and for data acquisition sent to a remote host by a 100 Mb Ethernet link.

The simulator used a blockset called RTeDRIVE, designed by OPAL-RT, which uses interpolation techniques to solve under sampling problems of the PWM

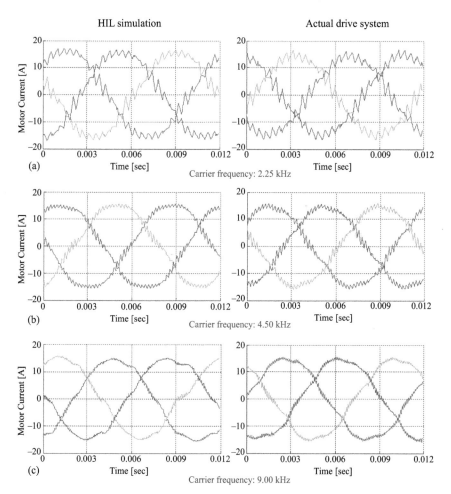

Figure 10.16   Effect of PWM carrier frequency

waveform by the RTS. As can be seen in Figure 10.16, HIL simulations closely match actual system results despite a 10-μs sample time and a nominal 9 kHz PWM frequency. Furthermore, one can observe that the current ripple amplitude decreases when PWM frequency is increased, just like in a real drive system.

One must note that today, such HIL simulation can be done with a time step of less than 1 μs using a complex finite element motor model implemented on a single FPGA chip as discussed in next sections.

10.8.2   FPGA modeling of SRM and PMSM motor drives

Recent advances in FPGA technologies now enable the entire simulation of motor drives on FPGA chips. In [26], an FPGA implementation of a switched reluctance motor (SRM) drive and an H-bridge Buck-Boost converter targeted for HIL testing of modern SRM controllers.

These FPGA models, shown in Figure 10.17, allow the HIL simulation of SRM drive and boost converter with switching frequencies in the 50–100 kHz range because of the very high sampling rate of the FPGA. The models are also integrated into the RT-LAB real-time environment and directly linked with the simulator I/Os, providing ultra-low HIL gate-in-to-current-out latency, suitable for testing motor controllers with ultra-low latency requirements.

Similar FPGA models have also been designed for PMSM drive [27]. This type of model is used to test hybrid vehicle drivelines, similar to the Prius car, composed of 2 PMSMs, depicted in Figure 10.18. These PMSM models include finite-element analysis (FEA) data and are sufficiently accurate to replace the real drive-line as tests made by the OEM show, as depicted in Figure 10.19.

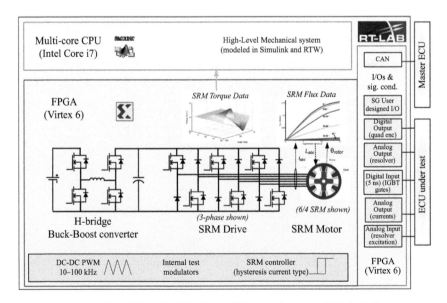

Figure 10.17   FPGA-based SRM drive HIL simulator with DC–DC converter

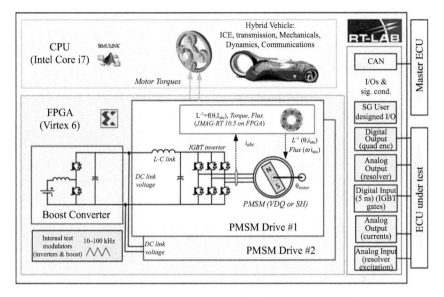

Figure 10.18   Dual-PMSM drive with boost converter of FPGA

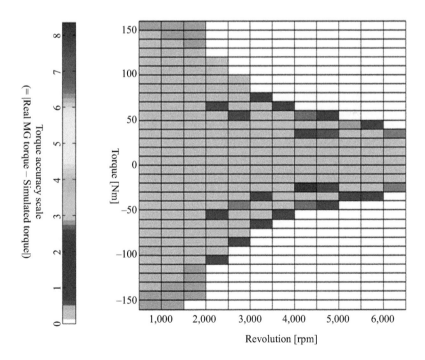

Figure 10.19   FEA model accuracy check with real driveline

10.9   Conclusion

This chapter has briefly presented various industrial applications of real-time simulation in the fields of power systems, motor drives, avionics, and robotics. A brief description of the model based design methodology was also presented together with a discussion of the main challenges encountered in the design of RTSs. It was demonstrated that the most complex applications found during the integration of very complex MMC HVDC grids can now be simulated in real-time to test actual control equipment using simulators based on standard computer systems.

As modern engineering projects become more complex, often with tight budgets and shortened development times, simulation technologies are becoming increasingly crucial to their success. It is believed that modern engineering curricula would benefit from the inclusion of real-time simulation technology courses because of widespread usage of simulation technology, both by industry and by researchers.

References

[1]   R. Kuffel, J. Geisbrecht, T. Maguire, R. P. Wierckx, and P. G. McLaren, "RTDS-a fully digital power system simulator operating in real time," in: WESCANEX – IEEE Conference on Communications, Power, and Computing, Winnipeg, MA, Canada, May 15–16, 1995.

[2]   R. Gagnon, G. Turmel, C. Larose, J. Brochu, G. Sybille, and M. Fecteau, "Large-scale real-time simulation of wind power plants into Hydro-Québec power system," in: Ninth International Workshop on Large-scale Integration of Wind Power into Power Systems as well as on Transmission Networks for Offshore Wind Plants, Quebec City, QC, Canada, October 18–19, 2010.

[3]   C. Dufour, J. Mahseredjian, J. Bélanger, and J. L. Naredo, "An advanced real-time electro-magnetic simulator for power systems with a simultaneous state-space nodal solver," in: *IEEE/PES T&D - Latin America*, São Paulo, Brazil, November 8–10, 2010.

[4]   O. Devaux, L. Levacher, and O. Huet, "An advanced and powerful real-time digital transient network analyser," *IEEE Trans. Power Delivery*, vol. 13, no. 2, pp. 421–426, 1998.

[5]   C. Dufour, T. Ould Bachir, L.-A, Grégoire, and J. Bélanger, "Real-time simulation of power electronic systems and devices," in: F. Vasca and L. Iannelli (Eds.), *Chapter 15 of the book Dynamics and Control of Switched Electronic Systems: Advanced Perspectives for Modeling, Simulation and Control of Power Converters*, Springer Series on Advances in Industrial Control, Springer, New York, NY, 2012. ISBN 978-1-4471-2885-4

[6]   C. Dufour, S. Cense, T. Ould Bachir, L.-A. Grégoire, and J. Bélanger, "General-purpose reconfigurable low-latency electric circuit and drive solver on FPGA," in: *38th IEEE Industrial Electronics Society (IECON)*, Montréal, QC, Canada, October 25–28, 2012.

[7]   C. Wang, W. Li, and J. Belanger, "Real-time and faster-than-real-time simulation of Modular Multilevel Converters using standard multi-core CPU

and FPGA chips," in: *39th IEEE Industrial Electronics Society (IECON)*, Vienna, Austria, November 10–13, 2013.

[8]   C. Dufour, J. Mahseredjian, and J. Bélanger, "A combined state-space nodal method for the simulation of power system transients," *IEEE Trans. Power Delivery*, vol. 26, no. 2, pp. 928–935, 2011.

[9]   C. Dufour and J. Bélanger, "On the use of real-time simulation technology in smart grid research and development," in: *IEEE Energy Conversion Congress and Exposition (ECCE)*, Denver, CO, USA, September 15–19, 2013.

[10] O. Tremblay, M. Fecteau, and P. Prud'homme, "Precise algorithm for non-linear elements in large-scale real-time simulator," in: *CIGRÉ Conference*, Montréal, QC, Canada, September 24–26, 2013.

[11] C. Dufour and O. Tremblay, "Iterative algorithms of surge arrester for real-time simulators," in: 18th Power Systems Computation Conference (PSCC), Wroclaw, Poland, August 18–22, 2014 (Accepted paper).

[12] C. Dufour, W. Li, X. Xiao, J.-N. Paquin, and J. Bélanger, "Fault studies of MMC-HVDC links using FPGA and CPU on a real-time simulator with iteration capability", in: *11th International Conference on Compatibility, Power Electronics and Power Engineering* (IEEE CPE-POWERENG 2017), Cadiz, Spain, April 4–6, 2017.

[13] A. Chandra, A. K. Pradhan, and A. K. Sinha, "PMU based real time power system state estimation using ePHASORsim," in: *2016 National Power Systems Conference (NPSC)*, Bhubaneswar, 2016, pp. 1–6. doi: 10.1109/NPSC.2016.7858967.

[14] E. Hogan, E. Cotilla-Sanchez, M. Halappanavar, et al., "Towards effective clustering techniques for the analysis of electric power grids," in: 3rd International Workshop on High Performance Computing, Networking and Analytics for the Power Grid (HiPCNA-PG), Denver, CO, USA, November 22, 2013.

[15] F. Pruvost, T. Cadeau, P. Laurent, F. Magoulès, F-X. Bouchez, and B. Haut, "Numerical accelerations for power systems transient stability simulations," in: 17th Power System Computation Conference (PSCC), Stockholm, Sweden, August 22–26, 2011.

[16] A. Viehweider, G. Lauss, and F. Lehfuss, "Stabilization of power hardware-in-the-loop simulations of electric energy systems," *Simul. Model. Pract. Theory*, vol. 19, no. 7, pp. 1699–1708, 2011.

[17] M. Hong, Y. Miura, T. Ise, Y. Sato, T. Momose, and C. Dufour, "Stability and accuracy analysis of power hardware-in-the-loop simulation of inductor coupled systems," *IEEJ Trans. Ind. Appl.*, vol. 130, no. 7, pp. 902–912, 2010.

[18] W. Ren, M. Steurer, and T.L. Baldwin, "Improve the stability and the accuracy of power hardware-in-the-loop simulation by selecting appropriate interface algorithm," *IEEE Trans. Ind. Appl.*, vol. 44, no. 4, pp. 1286–1294, 2008.

[19] D. Ocnasu, S. Bacha, I. Munteanu, and C. Dufour, "Real-time power-hardware-in-the-loop facility for shunt and serial power electronics benchmarking," in: 13th European Conference on Power Electronics and Applications (EPE), Barcelona, Spain, September 8–10, 2009.

[20] B. Vahidi, M. Ghorat, and E. Goudarzi, "Overvoltage calculation on bam substation by Monte Carlo method with accurate substation components modeling," in: IEEE PowerTech Conference, Lausanne, Switzerland, July 1–5, 2007.

[21] A. Lesnicar and R. Marquardt, "An innovative modular multilevel converter topology suitable for a wide power range," in: IEEE Power Tech Conference, Bologna, Italy, June 2003.

[22] L.-A. Grégoire, Wei Li, J. Bélanger, and L Snider, "Validation of a 60-level modular multilevel converter model – overview of offline and real-time HIL testing and results," in: *International Power Systems Transients (IPST)*, Delft, The Netherlands, July 2011.

[23] H. Saad, C. Dufour, J. Mahseredjian, S. Dennetière, and S. Nguefeu, "Real time simulation of MMCs using the state-space nodal approach," in: *International Power Systems Transients (IPST)*, Vancouver, BC, Canada, July 18–20, 2013.

[24] X. Li, Y. Liu,Y. Zhu ,T. Hu, C. Liu, and L.Chen, "Real-time simulation of dynamic performance of multi-infeed UHVDC transmission system to be connected to North China power grid before 2015," in: *Power System Technology*, 2011-08.

[25] M. Harakawa, H. Yamasaki, T. Nagano, S. Abourida, C. Dufour, and J. Bélanger, "Real-time simulation of a complete PMSM drive at 10 us time step," in: International Power Electronics Conference (IPEC), Niigata, Japan, April 4–8, 2005.

[26] C. Dufour, S. Cense, and J. Bélanger, "FPGA-based switched reluctance motor drive and DC–DC converter models for high-bandwidth HIL real-time simulator," in: 15th European Conference on Power Electronics and Applications (EPE-ECCE Europe), Lille, France, September 3–5, 2013.

[27] C. Dufour, S. Cense, T. Yamada, R. Imamura, and J. Bélanger, "FPGA permanent magnet synchronous motor floating-point models with variable-DQ and spatial harmonic finite-element analysis solvers," in: 15th International Power Electronics and Motion Control Conference (EPE-PEMC), Novi Sad, Serbia, September 4–6, 2012.

[28] J.N. Paquin, J. Bélanger, L.A. Snider, C. Pirolli, and W. Li, "Monte-Carlo study on a large-scale power system model in real-time using eMEGAsim," in: *IEEE Energy Conversion Congress and Exposition (ECCE)*, San Jose, CA, USA, September 20–24, 2009.

[29] J.C. Piedboeuf, F. Aghili, M. Doyon, Y. Gonthier, E. Martin, and W.H. Zhu, "Emulation of space robot through hardware-in-the-loop simulation", in: 6th International Symposium on Artificial Intelligence and Robotics & Automation in Space (i-SAIRAS), Canadian Space Agency, St-Hubert, QC, Canada, June 18–22, 2001.

[30] É. Martin, A. Lussier-Desbiens, T. Laliberté, and C. M. Gosselin, "SARAH hand used for space operations on STVF robot," in: International Conference on Intelligent Manipulation and Grasping, Genoa, Italy, July 1–2, 2004.

Chapter 11

Octsim/a solver for dynamic system simulation

Antonello Monti[1,2], Nika Khosravi[1], Martina Joševski[3] and Zhiyu Pan[1]

11.1 Introduction

When it comes to using a free alternative to MATLAB® for teaching, learning, or research, one has several options to choose from. Scilab, FreeMat, and GNU Octave, for example, are powerful open source numerical computing packages that share many of MATLAB's features. According to [1], all three packages show similar numerical outcomes.

Scilab has been developed by the French National Research Institution (INRIA) in 1990. Although it is intended to allow MATLAB users to exploit Scilab packages by utilizing a code translation tool, Scilab does not *per se* provide high-level syntax compatibility with MATLAB [1]. That way, using Scilab instead of MATLAB would require the user to learn the corresponding Scilab commands. Compared to MATLAB and Octave, Scilab also exhibits computational limitations in solving large-scale systems of linear equations [1]. Similar to Simulink® in the MATLAB software package, Scilab comes along with Xcos, that is, a block-based modeling and simulation tool.

FreeMat has been designed as free of charge alternative to MATLAB which is highly compatible with the MATLAB syntax. In 2013, the inventor Samit Basu claimed that FreeMat has a MATLAB command coverage of approximately 95% [2]. Along the lines of [1], though, FreeMat lacks a conjugate gradient function and has limited three-dimensional graphical visualization capabilities. Moreover, since 2013, FreeMat has not been updated any further [3]. As opposed to Scilab, FreeMat does not come along with a Simulink equivalent.

GNU Octave is a matrix-based high-level programming language designed primarily for numerical calculations [4]. The GNU General Public License Project governs the distribution of Octave, which is a free and open-source software [4,5]. Octave is highly compatible with MATLAB [4], features a higher coverage of MATLAB commands than FreeMat [1], and is more powerful, e.g., when it comes to the solution of large-scale problems. Linear algebra, the numerical solution of linear and nonlinear regression problems, numerical optimization problems, the solution of integrals as well

[1]Institute for Automation of Complex Power Systems, RWTH Aachen University, Germany
[2]Fraunhofer FIT Center for Digital Energy, Germany
[3]Eaton Industries GmbH, Germany

as the ability of plotting and graphical visualization are only some examples of Octave's characteristics and capabilities, which makes it an attractive tool for teaching and research [6,7]. Based on [1,5,8], one can conclude that Octave is most compatible and comparable with MATLAB due to its numerical abilities and the similarity of its syntax – only a Simulink-like environment to model and simulate dynamical systems is missing.

To close this gap, the authors have developed *Octsim* – an Octave-based solver for dynamical system simulation, integrated with Jupyter Notebook [9] which also serves as a graphical frontend. Octsim can be exploited to model and simulate continuous-time and discrete-time systems as well as a combination of both (in the remainder referred to as *hybrid*). Due to its practical teaching features and learning benefits [2], Jupyter Notebook, also known as a computational notebook, is a perfect environment for integrating and testing Octsim. While Octsim has been developed to be used integrated in a Jupyter notebook, it can also be used as a text-based simulator in a standard Octave program or script.

This chapter describes the utilization of Octsim along the lines of the *System-Theory II* course at the Institute for Automation of Complex Power Systems at RWTH Aachen University. The authors will outline that Octsim constitutes a reliable alternative to Simulink to model simulate the behavior of dynamical systems – a feature that has been missing until now. In the remainder, the chapter outlines the design, implementation, and validation of Octsim along with some illustrative examples.

This example of application also demonstrates how an object-oriented library can be easily created to support a general purpose solver that can be easily extended both for circuit-based and block diagram-based simulation. The software is available in a github repository and the authors welcome contributions and extensions.

11.2 Solver description

In the following section, the solver structure and functionalities of Octsim are described in detail.

11.3 Solver structure

Donald Knuth first presented the Literate Programming paradigm in 1983 to improve understanding in the phase of comprehending code written by another programmer, as well as to provide a solid basis for teaching and learning [10,11]. The Literate Programming model proposes the concept of incorporating code fragments and natural language text to demonstrate to other humans what we expect the computer to do and the reasoning behind the code [9,12].

The concept of literate programming is commonly used in computational notebooks, such as the renowned Jupyter Notebook [9]. Jupyter Notebook is a cross-platform, web-based collaborative computing environment for document creation [3] and it has three different kinds of cells: code, markdown, and raw. There are various language-specific computing engines, the so-called *kernels*, for the execution of code cells [9]. The key focus of this chapter is the usage of the Octave kernel to simulate dynamical systems through Octsim.

RWTH Aachen University has established the so-called RWTHjupyter Hub, which is a server that allows a vast number of users to access Octsim, Jupyter Notebook, and other computing environments without the need to install any software packages [13,14]. That said, RWTHjupyter can very conveniently be leveraged for lectures or exercises in an *ad hoc* fashion as students only need to access the respective webpage (via http) with their laptop, smartphone or tablet.

RWTHjupyter consists of 7 Dell PowerEdge R740xd servers with the following configurations:

- Dual socket systems: $2\times$ 16C / 32T Xeon Gold 5218 2.3 GHz.
- Redundant dual 10 GigE network connection.
- 768 GB DDR4 RAM/node (total 5.3 TB).
- 100 TB SSD hyperconverged SSD storage in Ceph.
- Additionally, one node is equipped with two NVIDIA Tesla T4 GPGPUs for programming with CUDA, Tensorflow, PyTorch, or other machine learning frameworks [15].

Along the lines of Figure 11.1, RWTHjupyter contains a series of profiles, each being relevant to an RWTH Aachen University course. The System-Theory II

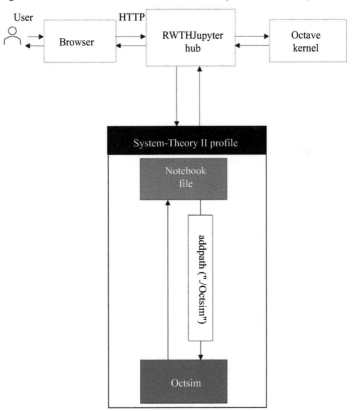

Figure 11.1   Octsim structure integrated with Jupyter Notebook

profile includes the course's notebook file as well as the Octsim library. Simply adding Octsim to the workspace and the Octsim directory to the simulation route is all that is required to get started with Octsim, that is:

```
[
addpath("./octsim")
]:
```

or

```
[
SetSimulationEnvironment;
]:
```

Technically, the Octave kernel receives and executes code fragments which are defined in the code cells of Jupyter Notebook. After the Octave kernel has executed the code, it returns the results to the Jupyter Notebook [13] and as such finally to the user. Figure 11.1 illustrates the architecture of Jupyter Notebook integrated with Octsim.

11.4   Solver functionalities

Octsim has two types of classes as illustrated in Figure 11.2: (1) the first is specifically constructed for control system simulation and (2) the second is useful for

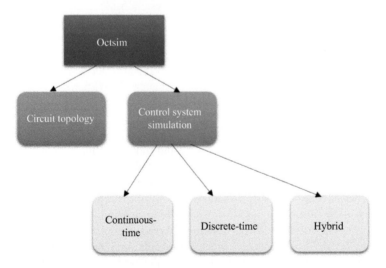

Figure 11.2   Octsim classes for (1) control system simulation and (2) electric circuit simulation

specifying a circuit topology. For both applications, the models can be directly defined in a notebook file, which is linked to Octsim as discussed previously.

For control system simulation (Type 1), the system of interest can conveniently be described through Octsim functions which are utilized in the notebook's code cells. **StateSpace**, **DTStatespace**, and **HybridSystem** constitute the respective classes for continuous-time, discrete-time, and hybrid systems, respectively. After running the simulation by executing the commands in the code cells (using the Octave kernel), the simulation outcomes can be observed in the Jupyter Notebook [16].

To analyze the behavior of a circuit (Type 2), the circuit topology and parameters need to be defined in a Jupyter Notebook. To simulate the components of the circuit, Octsim provides several classes such as **ACCurrentSource**, **Diode**, and **Resistance**. Again, the simulation is run by executing the corresponding code cell, and its results can thereafter be plotted for further study.

11.5   Solver implementation and validation

11.5.1   Implementation details

The Octave programming language is used to develop the solver and its components, thereby exploiting the concept of object-oriented programming.

According to [17,18], the main object-oriented concepts are related to Octsim in following aspect:

1. **Data abstraction:** The exact implementation of each Octsim class is stored as an encapsulated code. By decoupling the behavior from implementation [19], Octsim offers the user to easily model the system he wants with just knowing the abstract description of each class [18].

   By saying the abstract description, we mean, by just knowing which or how many inputs we require for creating an object. For example, if a user wishes to construct a Kalman filter, he needs to grasp the abstracted description of the Kalman filter, which is written in the user manual. Hereafter, he knows which or how many input parameters he requires to implement the desired Kalman Filter with.

2. **Homogeneity:** Octsim provides the user with a variety of objects as well as procedures for manipulating the objects and modeling the behavior of a system. Moreover, Octsim is a fully object-oriented system and therefor offers a high level of homogeneity. This means that the solver does not see any difference between classes or procedures and all of them are objects [17].

3. **Inheritance:** With an object-oriented implementation, Octsim takes advantage of class inheritance to build the solver's hierarchy. Inheritance is an important property in object-oriented programming because it allows new subclasses to inherit variables and methods from the so-called base- or superclasses and to extend these by its own set of features and capabilities [20,21]. That said, an object-oriented implementation of Octsim allows

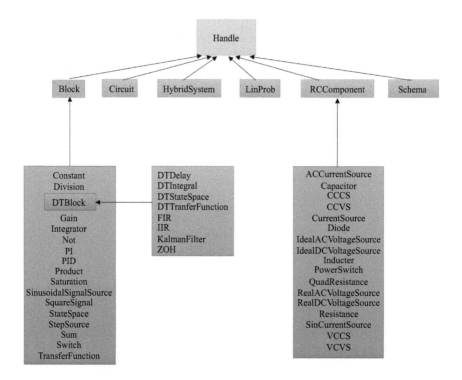

Figure 11.3   Overview of Octsim class hierarchy

Octsim developers to easily extend and more importantly to reuse class features and functionalities when creating new, derived classes for simulation purposes [22].

4. **Independence:** Object "independence" is the notion that each object is ultimately responsible for itself [23]. This means objects can only be manipulated via their methods. The Octsim solver follows this principle, which each of the defined class is independent. The Octsim classes can only operate the method in the class, which is hidden from other classes.

Classes implemented in Octsim are proper descendants of the abstract Octave class handle due to their direct or indirect inheritance from this base class [24]. Moreover, instances of a handle (sub-)class are handle objects [25].

By virtue of Figure 11.3, the Octsim class hierarchy features three layers of inheritance. The first layer of inheritance is constituted by the classes Block, Circuit, HybridSystem, LinProb, RCComponent, and Schema (shown in green), which are all derived from the abstract class handle. In the second layer, two groups of classes are derived from the classes Block and RCComponent (shown in blue), which inherit their methods and properties. The final layer (shown in red) is given by classes that are directly derived from DTBlock.

11.5.2 Comparison with Simulink

To evaluate the validity of the Octsim implementation, simulation results that are obtained in Octsim for an open-loop and closed-loop system in state space form, i.e.,

$$\dot{x} = Ax + Bu$$

$$y = Cx + Du$$

are compared to the respective simulation results in Simulink. The system matrices of the **open-loop system** are given as:

$$A = \begin{bmatrix} 1 & 1 \\ -2 & -1 \end{bmatrix}, B = \begin{bmatrix} 0 \\ 1 \end{bmatrix}, C = \begin{bmatrix} 1 & 0 \\ 0 & 1 \end{bmatrix}, D = \begin{bmatrix} 0 & 0 \end{bmatrix}$$

with the initial condition at the initial time t_0:

$$x(t_0) = \begin{bmatrix} 2 & 2 \end{bmatrix}^T$$

The Simulink implementation of the system is depicted in Figure 11.4. Through the switch (top left), the inputs of the dynamical system are set to zero. There is also a state-feedback loop which however does not affect the system. For that reason, Figure 11.4 represents the open-loop system.

The system outputs y are observed through the scope on the top right and their values over time are illustrated in Figure 11.5. The corresponding Octsim outputs are given in Figure 11.6. To focus on the comparison of simulation results, the Octsim implementation is not covered in this section but is thoroughly discussed at a later stage. For both, Octsim and Simulink, a sample time of 0.01 seconds is applied and the same system matrices and initial conditions are used. The current vision of Octsim supplies only fixed-step solver.

By virtue of Figures 11.5 and 11.6, it can be recognized that Octsim produces the same results as Simulink.

In the following example, we turn the switch (top left) in Figure 11.4 from the upper to the lower position, thus closing the state feedback loop. That said,

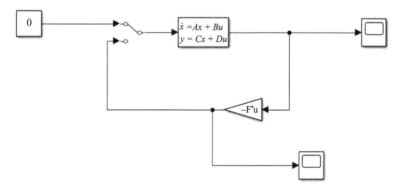

Figure 11.4   Simulation model of the open-loop system in Simulink

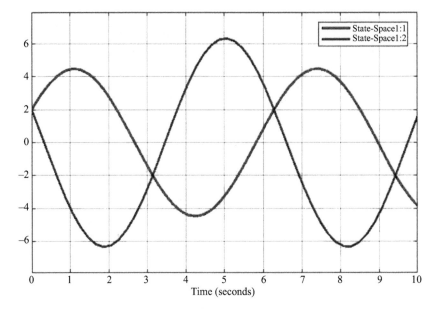

Figure 11.5   Simulink: output signal y for the open-loop system

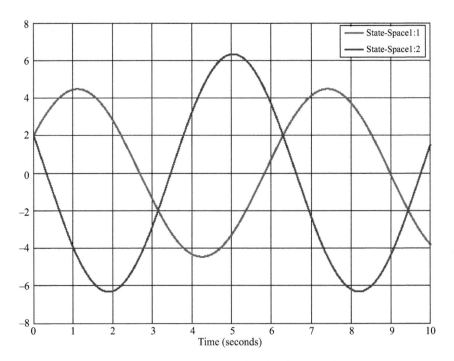

Figure 11.6   Octsim: output signal y for the open-loop system

Figure 11.7 illustrates the **closed-loop control system**. For comparison, the Simulink system outputs y of the top right scope in Figure 11.7 are depicted in Figure 11.8 while the respective Octsim system outputs y are shown in Figure 11.9. Similar to the previous example, a sample time of 0.01 seconds, and the same system matrices and initial conditions are applied in both cases. For the Octsim implementation of the closed-loop system simulation, the interested reader is again referred to the subsequent section.

Similar to the open-loop system, it is evident that the results of both simulation environments are almost indistinguishable.

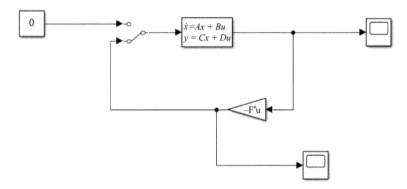

Figure 11.7   Closed-loop system in Simulink

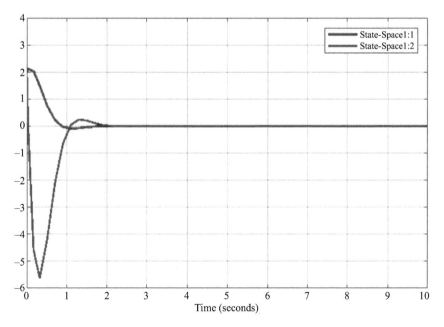

Figure 11.8   Simulink: output signal y for the closed-loop system

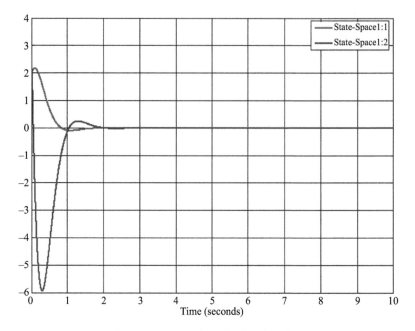

Figure 11.9   Octsim: output signal y for the closed-loop system

It can be concluded that Octsim, based on the Octave kernel and embedded in Jupyter Notebook, is a suitable alternative to simulate such kind of dynamical systems without the need for a MATLAB/Simulink® installation.

11.5.3   Octsim code examples

While the previous section has predominantly focused on a comparison of simulation results, this section gives an overview of how to implement the simulation of a control system or an electrical circuit using Octim.

11.5.4   Control system simulation

To simulate a control system with Octsim, blocks in Simulink can be associated with an Octsim class (see Figure 11.4). Let us contemplate the example of the **open-loop system** in Figure 11.10. The Simulink Constant block corresponds to the Constant class, the dynamical system in state-space form to the StateSpace class and the gain block to the Gain class. An instance of each of these classes, subsequently referred to as object, can be created by calling the respective constructor [26].

That said, every Octsim object corresponds to exactly one simulation component, which exhibits a unique identifier (numbering starts at 1). Getting back to the example in Figure 11.4, a data flow identifier (scalar or vector) must be associated with each input and output. In the example at hand, these data flow identifiers are 1 for the constant input, [2 3] for the output of the state space model and 4 for the control input resulting from the state feedback control law.

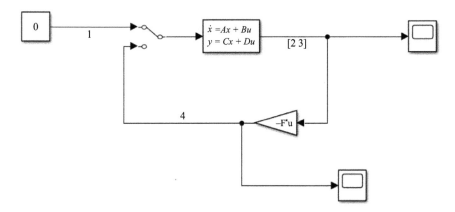

Figure 11.10   Open-loop system with data flow in Simulink

Data flow identifier 1 refers to the output of the Constant element, which thereafter serves as an input to the StateSpace object. The Constant constructor is defined as follows:

```
[
Constant (1.argument=output flow, 2.argument=value)
]:
```

The Constant is then instantiated as the first component c1{1} of the Octsim simulation, holding the data flow identifier 1 and a constant output of 0, i.e.,

```
[
c1{1} = Constant (1,0);
]:
```

The StateSpace constructor expects the data flow identifier of the input and output signals as the first and the second argument, respectively. The following arguments refer to the system matrices of the dynamical system in state-space representation along with the initial condition as last argument.

```
[
StateSpace(1.argument=input,          2.argument=output,
          3.argument=system matrix, 4.argument=input
matrix,
          5.argument=output matrix,
6.argument=feedforward matrix,
          7.argument=initial condition)
]:
```

The StateSpace object in Octsim is created by assigning the data flow identifier 1 (output of the constant object c1{1}) to the input of the state space object. Its outputs are uniquely identified through the vector [2 3] (two outputs). The remaining inputs are the system matrices (A,B,C,D) and the initial condition xo, which need to be defined in the Jupyter Notebook. The StateSpace object is the second element c1{2} of the Octsim simulation.

```
[
c1{2} = StateSpace(1,[2 3],A,B,C,D,xo);
]:
```

As the switch on the top left is in the upper position, the state feedback law which is described by the Gain block does not have any influence on the input of the state space model.

To contemplate the **closed-loop system**, the switch has to be adjusted to the lower position. Then, the Gain class in Octsim becomes relevant. Its constructor expects the data flow identifier for the input and output signals as the first two arguments followed by the scalar of vector type multiplier, that is,

The data flow number four (4) is the performance of the Gain constructor:

```
[
Gain(1.argument=input, 2.argument=output,
3.argument=value)
]:
```

When creating the Gain object as the third element c1{3} of the Octsim simulation, the data flow identifiers [2 3] of the StateSpace outputs are defined as inputs to the Gain object while its outputs are associated with the identifier 4. The last argument constitutes the state feedback control law -F.

```
[
c1{3} = Gain([2 3],4,-F);
]:
```

To finally close the loop, the inputs of the StateSpace object need to be modified. To this end, the StateSpace object is created by using the data flow identifier 4 (Gain) instead of 1 (Constant) as input identifier.

```
[
c1{2} = StateSpace(4,[2 3],A,B,C,D,xo);
]:
```

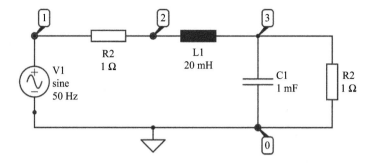

Figure 11.11   Circuit diagram

11.5.5   *Electric circuit simulation*

The most reliable way for defining a circuit topology with the current version of Octsim is to code the topology using Octsim's classes. The basic concept is based on three steps:

1.   Define the circuit in terms of number of nodes.
2.   Create the component with node information.
3.   Add the component to the circuit.

As an example, the following circuit shown in Figure 11.11 is simulated:
An object of the circuit class with three nodes is created:

```
[
sc = Circuit(3);
]:
```

The angular frequency is modified based on the frequency of an ideal AC voltage source:

```
[
om = 2*pi*50;
]:
```

The ideal AC voltage source with the amplitude of 100 V, defined angular frequency of om and the phase of $\frac{2\pi}{3}$ is created:

```
[
vs = IdealACVoltageSource(1,0,100,om,2*pi/3);
sc.AddComponent(vs);
]:
```

A resistor *r*1 from node 1 to node 2 with the value of 1 Ω is added to the simulation using sc.AddComponent(*r*1). In the same way, the inductor l1 from node 2 to node 3 with 0.02 H, the capacitor c1 from node 3 to node 0 with 0.001 F, and the resistor *r*2 from node 3 to node 0 with 1 Ω are created and added to the circuit class.

```
[
r1 = Resistance(1,2,1);
sc.AddComponent(r1);
]:
```

```
[
l1 = Inductor(2,3,0.02);
sc.AddComponent(l1);
]:
```

```
[
c1 = Capacitor(3,0,0.001);
sc.AddComponent(c1);
]:
```

```
[
r2 = Resistance(3,0,1);
sc.AddComponent(r2);
]:
```

11.6   Example for hybrid system (buck converter with voltage control)

Here the circuit shown in Figure 11.12 is simulated:

At first, the following values are given to the parameters of a hybrid system:

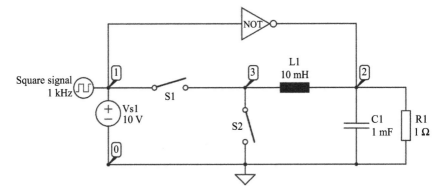

Figure 11.12   Circuit diagram

```
[
tini = 0;
tfinal = 0.1;
dt = 0.00001;
nflows = 2;
nnode = 3;
maxn = 20;
toll = 0.0001;
]:
```

And an object of the HybridSystem class is created:

```
[
hy = HybridSystem(nnode,nflows,tini,tfinal,dt,maxn,
    toll);
]:
```

As next step, a DC voltage source with the voltage of 10 V and the inner resistance of 0.01 Ω is created:

```
[
vs1 = RealDCVoltageSource(1,0,10,0.01);
hy.AddComponent2Network(vs1);
]:
```

As showed in the provides example, the inductor *l*1 from node 3 to node 2 with 0.01 H, the capacitor *c*1 from node 2 to node 0 with 0.001 F, and the resistor *r*2 from node 2 to node 0 with 1 Ω are created and added to the circuit class in the same way:

```
[
L1 = Inductor(3,2,0.01);
hy.AddComponent2Network(L1);
]:
```

```
[
c1 = Capacitor(2,0,0.001);
hy.AddComponent2Network(C1);
]:
```

```
[
R1 = Resistance(2,0,1);
hy.AddComponent2Network(R1);
]:
```

Afterward, two switches and a Not component are created according to Figure 11.12 circuit diagram:

```
[
% Switch 1
S1 = PowerSwitch(1,3,1,0.0001,1e4);
hy.AddComponent2Network(S1);
% Switch 2
S2 = PowerSwitch(3,0,2,0.0001,1e4);
hy.AddComponent2Network(S2);
]:
```

```
[
N1 = Not(1,2); hy.AddComponent2Schema(N1);
]:
```

Last, a Square wave is generated and added to the HybridSystem class:

```
[
SW1 = SquareSignal(1,1,0,1000,0.5);
hy.AddComponent2Schema(SW1);
]:
```

Finally, all the component is added and we can simulate the results:

```
[
hy.Init();
p=1;
while hy.Step()
   time(p) = hy.GetTime();
   out1(p) = hy.GetFlow(2);
   out2(p) = hy.GetFlow(1);
   out3(p) = hy.GetNode(2);
   out4(p) = hy.GetNode(1);
   p=p+1;
end
plot(time,out3);
]:
```

Figure 11.13 shows the current profile at the node 2. After 0.06 seconds, the current reaches the steady stable at 5A as expected.

11.7 Conclusion

The major motivation for the design of Octsim library has been to provide a practical, reliable, and free tool for Octave with Simulink-like capabilities that can be applied to academic purposes.

Octsim is primarily a code-based solver, but visual and graphical enhancements will be introduced in the future. We envision the Octsim library adding a graphical user interface to go along with its textual functionality. Additionally, we aim to develop a parser for XML files related to the Simulink model. This design step can significantly improve the interoperability of Octsim since it provides the user a possibility to build the graphical diagram in Simulink and run the solver in Octsim. Another limitation is the current vision of Octsim supplies only fixed-step solver. But the variable-step will be developed in the further vision. The last point is that Octsim executes the code from top to bottom, in order to execute the blocks in a correct order, the sequence of the blocks should be well defined. One possible solution to this

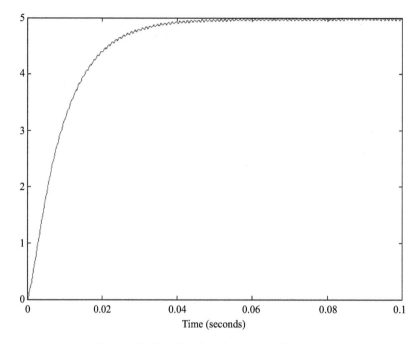

Figure 11.13   Simulated result of the node 2

is to employ the Petri nets that could determine the next state of the system, validate
if the system can be executed or identify if there is an algebraic loop.

11.8   User manual

- By adding the Octsim directory to the simulation path, the user gains access to
  all necessary functionalities. To begin the simulation, the user must
  configure the Octave Engine as follows:

```
[
% Set the Octave Engine to run the simulation
addpath("./Octsim");

% Alternative:
SetSimulationEnvironment;
]:
```

- The next step is to define the required simulation parameters such as initial time, final time, time step and the number of data flows in the schematic:

```
[
% Start time
tini = 0;

% End time
tfinal = 10;

% Time Step
dt = 0.01;

% Number of data flows in the schematic
nflows = 4;
]:
```

- Following that, an object of the schematic class may be generated. The Schema function constructs the simulation schematic by taking as inputs simulation parameters such as beginning time, final time, time step, and the number of data flows in the schematic:

```
[
% Instance of the simulation schematic
sc = Schema(tini,tfinal,dt,nflows);
]:
```

- The next step is to use the Octsim to generate objects of each class required for the simulation of a desired model (see Example 1). Table 11.1 is a list of Octsim classes and operations.
- The final step is to execute the schematic using the previously created schematic object and visualize the results.

Basic Example: A system is defined in the state space and the free evolution of the system is being simulated in open loop.

Figure 11.14. A basic example of diagram that can be simulated in Octsim

Table 11.1   Octsim classes

ACCurrentSource	ACCurrentSource(1. argument = positive terminal, 2. argument = negative terminal, 3. argument = amplitude, 4. argument= frequency, 5. Argument = Phase)
Block	Block( )
Capacitor	Capacitor(1. argument = positive terminal, 2. argument = negative terminal, 3. Argument = value of the capacitor)
CCCS	CCCS(1. argument = positive terminal, 2. argument= negative terminal, 3. argument = current at positive control port, 4. argument= current at negative control port, 5. argument= gain)
CCVS	CCVS(1. argument = positive terminal, 2. argument = negative terminal, 3. argument = positive terminal of controlling voltage, 4. argument = negative terminal of controlling voltage, 5. argument = gain)
Circuit	Circuit(number of nodes)
Constant	Constant(1. argument = output flow identifier, 2. argument = value)
CurrentSource	CurrentSource(1. argument = positive terminal, 2. argument = negative terminal, 3. argument = value)
Diode	Diode(1. argument = positive terminal, 2. argument = negative terminal, 3. argument = on-state resistance, 4. argument = off-state resistance)
	Diode(1. argument = positive terminal, 2. argument = negative terminal)
Division	Division(1. argument = input flow identifier #1, 2. argument = input flow identifier #2, 3. argument = output flow identifier)
DTBlock	DTBlock( ) – abstract class for block diagrams
DTDelay	DTDelay(1. argument = input flow identifier, 2. argument= output flow identifier, 3. argument = initial condition, 4. argument = sampling time)
DTIntegral	DTIntegral(1. argument = input flow identifier, 2. argument = output flow identifier, 3. argument = sampling time, 4. argument = initial condition)
DTStatespace	StateSpace(1. argument = input flow identifier, 2. argument = output flow identifier, 3. argument = system matrix, 4. argument = input matrix, 5. argument = output matrix, 6. argument = feedforward matrix, 7. argument = sampling time)
DTTransferfunction	DTTransferfunction(1. argument = input flow identifier, 2. argument = output flow identifier, 3. argument = numerator coefficients, 4. argument = denominator coefficients, 5. argument = sampling time)
FIR	FIR(1. argument = input flow identifier, 2. argument = output flow identifier, 3. argument = vector of coefficients of the filter, 4. argument = sampling time)
Gain	Gain(1. argument = input flow identifier, 2. argument = input flow identifier, 3. argument = gain)
HybridSystem	Hybridsystem(1. argument = number of nodes, 2. argument = number of flows, 3. argument = initial time, 4. argument = final time, 5. argument = time step, 6. argument = max iterations non-linear solver, 7. argument = tolerance non-linear solver)

Table 11.1 (*Continued*)

IdealACVoltageSource	IdealACVoltageSource(1. argument = positive terminal, 2. argument = negative terminal, 3. argument = value, 4. argument = angular frequency, 5. argument= phase)
IdealDCVoltageSource	IdealDCVoltageSource(1. argument = positive terminal, 2. argument = negative terminal, 3. argument = value)
IIR	IIR(1. argument = input flow identifier , 2. argument = output flow identifier, 3. argument = transfer function coefficients of numerator, 4. argument = transfer function coefficients of denominator, 5. argument = sampling time)
Inductor	Inductor(1. argument = positive terminal, 2. argument = negative terminal, 3. argument = value of the inductance)
Integrator	Integrator(1. argument = input flow identifier, 2. argument = output flow identifier, 3. argument = initial conditions)
KalmanFilter	KalmanFilter(1. argument = input flow identifier, 2. argument = output flow identifier, 3. arguments = number of inputs, 4. argument = number of measurements, 5. argument = sampling time, 6. arguments = system matrix, 7. arguments = input matrix, 8. arguments = output matrix, 9. arguments = feedforward matrix, 10. argument = covariance matrix R, 11. argument = initial guessed conditions, 12. argument = uncertainty on the initial conditions)
LinProb	LinProb(n) – class to manage linear problem solution
Not	Not(1. argument = input flow identifier, 2. argument = output flow identifier)
PI	PI(1. argument = input flow identifier, 2. argument = output flow identifier, 3. argument = proportional gain, 4. argument = integral gain, 5. argument = initial conditions)
PID	PID(1. argument= input flow identifier, 2. argument= output flow identifier, 3. argument= proportional gain, 4. argument= integral gain, 5. argument= derivative gain , 6. argument= initial conditions)
PowerSwitch	PowerSwitch(1. argument = positive terminal, 2. argument = negative terminal, 3. argument = control terminal, 4. argument = on-state resistance, 5. argument = off-state resistance)
Product	Product(1. argument = input flow identifier #1, 2. argument = input flow identifier #2, 3. argument = output flow identifier)
QuadResistance	QuadResistance(1. argument = positive terminal, 2. argument = negative terminal, 3. argument = value)
RCComponent	RCComponent(1. argument = output flow identifier, 2. argument = input flow identifier)
RealACVoltageSource	RealACVoltageSource(1. argument = positive terminal, 2. argument = negative terminal, 3. argument = value, 4. argument = internal resistance, 5. argument = angular frequency , 6. argument = phase)
RealDCVoltageSource	RealDCVoltageSource(1. argument = positive terminal, 2. argument = negative terminal, 3. argument = value voltage, 4. argument = internal resistance)

(Continues)

Table 11.1   (Continued)

Resistance	Resistance(1. argument = positive terminal, 2. argument = negative terminal, 3. argument= value)
Saturation	Saturation(1. argument = input flow identifier, 2. argument = output flow identifier, 3. argument = saturation min, 4. argument = saturation max)
Schema	Schema(1. argument = initial time, 2. argument = final time, 3. argument = time step, 4. argument = number of data flows)
SinCurrentSource	SinCurrentSource(1. argument = positive terminal, 2. argument = negative terminal, 3. argument = amplitude, 4. argument = radiant frequency, 5. argument = phase)
SinusoidalSignalSource	SinusoidalSignalSource(1. argument = output flow identifier, 2. argument = amplitude, 3. amplitude = angular frequency, 4. argument = phase)
SquareSignal	SquareSignal(1. argument = output flow identifier, 2. argument = high, 3. argument = low, 4. argument = frequency, 5. argument = duty cycle)
StateSpace	StateSpace(1. argument = input flow identifier, 2. argument = output flow identifier, 3. argument = system matrix, 4. argument = input matrix, 5. argument = output matrix, 6. argument = feedforward matrix, 7. argument = initial conditions)
StepSource	StepSource(1. argument = output flow identifier, 2. argument = start value, 3. argument = end value, 4. argument = step time)
Sum	Sum(1. argument = input flow identifier #1, 2. argument = input flow identifier #2, 3. argument = output flow identifier, 4. argument = first sign, 5. argument = second sign)
Switch	Switch(1. argument = input flow identifier #1, 2. argument = control terminal, 3. argument = input flow identifier #2, 4. argument = output flow identifier)
TransferFunction	TransferFunction(1. argument = input flow identifier, 2. argument = output flow identifier, 3. argument = numerator coefficients, 4. argument = denominator coefficients)
VCCS	VCCS(1. argument = positive terminal current source, 2. argument = negative terminal current source, 3. argument = voltage at positive control port , 4. argument = voltage at negative control port, 5. argument= gain)
VCVS	VCVS(1. argument = positive terminal, 2. argument = negative terminal, 3. argument = voltage at positive control port, 4. argument = voltage at negative control port, 5. argument = gain)
WhiteNoise	WhiteNoise(1. argument = output flow identifier, 2. argument = seed value)
ZOH	ZOH(1. argument = input flow identifier, 2. argument = output flow identifier, 3. argument = sampling time)

The system is described by the following state space matrices:

```
[
A = [1 1; -2 -1];
B = [0; 1];
C = [1 0];
D = 0;
]:
```

The Octave Engine is set to run the simulation:

```
[
% Set the Octave Engine to run the simulation
addpath("./Octsim");
]:
```

The simulation parameters are defined:

```
[
% Start time
tini = 0;

% End time
tfinal = 10;

% Time Step
dt = 0.01;

% Number of data flows in the schematic
nflows = 4;
]:
```

The initial conditions could also be defined:

```
[
% Initial Conditions
xo = [2; 2];
]:
```

Now an instance of the simulation schematic will be created:

```
[
% Instance of the simulation schematic
sc = Schema(tini,tfinal,dt,nflows);
]:
```

Figure 11.1 shows a state-space model with a constant input. Based on this, the objects are defined. The first argument of Constant function represents the output with a flow number, which is arbitrary starting from one. The second argument contains the desired value for the function.

In the StateSpace, the flow number of the input is assigned to the first argument and the second argument has the value of the output flow number.

The instance created from each class is assigned to a component, which is marked with a number starting from one. Later on by adding the components, the simulation could be run.

```
[
% List of components
c{1} = Constant(1,0);
c{2} = StateSpace(1,2,A,B,C,D,xo);
]:
```

Now all the system components are added.

```
[
% Add the system components
sc.AddListComponents(c1);
]:
```

The final step is to run the schematic and plot the results:

```
[
out = sc.Run([5]);
plot(out(1,:),out(2,:));
]:
```

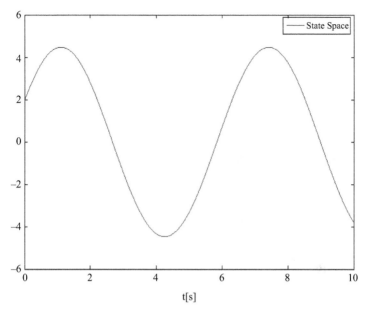

t[s]

References

[1] N. Sharma and M.K. Gobbert, "A comparative evaluation of MATLAB, Octave, FreeMat, and Scilab for research and teaching", *UMBC Faculty Collection*, 2010.

[2] http://freemat.sourceforge.net.

[3] D.E. Knuth, "Literate programming", *The Computer Journal*, Vol. 27, no. 2, pp. 97–111, 1984, DOI: 10.1093/comjnl/27.2.97.

[4] J.W. Eaton, "Octave: past, present and future", in *Proceedings of the 2nd International Workshop on Distributed Statistical Computing*, 2001.

[5] A. Pajankar, S. Chandu, *GNU Octave by Example: a Fast and Practical Approach to Learning GNU Octave*. New York, NY: Apress, 2020.

[6] J.W. Eaton, "GNU Octave and reproducible research", *Journal of Process Control*, Vol. 22, no. 8, pp. 1433–1438, 2012, DOI: 10.1016/j.jprocont.2012.04.006.

[7] J.S. Hansen, *GNU Octave: Beginner's Guide: Become a Proficient Octave User by Learning this High-Level Scientific Numerical Tool from the Ground Up*, Packt Publishing Ltd, Birmingham, 2011.

[8] J.W. Eaton, 1997, "A high-level interactive language for numerical computations", Edition 3 for Octave version 2.1.x.

[9] N. Ramsey, "Literate programming simplified", *IEEE Software*, Vol. 11, no. 5, pp. 97–105, 1994, DOI: 10.1109/52.311070.

[10] J.M. Perkel, "Why Jupyter is data scientists' computational notebook of choice", *Nature*, Vol. 563, no. 7732, p. 145, 2018. +. *Gale OneFile: Health and Medicine*, link.gale.com/apps/doc/A573082717/HRCA?u=anon~6d6512 f5&sid=HRCA&xid=dfeccfed. Accessed 11 May 2021.

[11] D. Cordes and M. Brown, "The literate-programming paradigm", *Computer*, Vol. 24, no. 6, pp. 52–61, 1991. DOI: 10.1109/2.86838.

[12]   J.F. Pimentel, L. Murta, V. Braganholo, and J. Freire, "A large-scale study about quality and reproducibility of Jupyter Notebooks", in *2019 IEEE/ACM 16th International Conference on Mining Software Repositories (MSR)*, Montreal, QC, Canada, 5/25/2019–5/31/2019: IEEE, pp. 507–517.

[13]   D. Geleßus and M. Leuschel, "ProB and Jupyter for logic, set theory, theoretical computer science and formal methods", in A. Raschke, D. Méry, F. Houdek (Eds.), *Rigorous State-Based Methods*, Lecture Notes in Computer Science, Vol. 12071, Springer International Publishing, Cham, 2020, pp. 248–254.

[14]   https://blog.rwth-aachen.de/itc/2020/10/28/rwthjupyter/.

[15]   Project Jupyter – A Multi-User Version of the Notebook Designed for Companies, Classrooms and Research Labs, https://jupyter.org/hub.

[16]   M. Mirz, S. Vogel, G. Reinke, and A. Monti, "DPsim—a dynamic phasor real-time simulator for power systems", *SoftwareX*, Vol. 10, p. 100253, 2019, DOI: 10.1016/j.softx.2019.100253.

[17]   European Organization for Nuclear Research, Geneva (Switzerland) (Apr 1987). 1986 CERN School of Computing (CERN-87-04). Verkerk, C. (Ed.). European Organization for Nuclear Research (CERN).

[18]   H. Mössenböck, *Object-Oriented Programming in Oberon-2*, Springer Science & Business Media, Berlin, 2012.

[19]   B. Liskov, "Keynote address-data abstraction and hierarchy", *Addendum to the Proceedings on Object-Oriented Programming Systems, Languages and Applications* (Addendum), 1987.

[20]   R. Lafore, *Object-Oriented Programming in C++*, Pearson Education, London, 1997.

[21]   I.D. Craig, *Object-Oriented Programming Languages: Interpretation*, Springer, New York, NY, 2007.

[22]   P. Wegner, "Concepts and paradigms of object-oriented programming", *SIGPLAN OOPS Messenger*, Vol. 1, no. 1, pp. 7–87, 1990, https://doi.org/10.1145/382192.383004.

[23]   S. Nygaard and K. Gjessing, *Proceedings of the European Conference on Object-Oriented Programming*, Lecture Notes in Computer Science, Springer, New York, NY, 1988.

[24]   B. Meyer, *Object-Oriented Software Construction*, 2nd ed., Prentice Hall, Upper Saddle River, NJ, 1997.

[25]   https://la.mathworks.com/help/matlab/ref/handle-class.html

[26]   MATLAB® Object-Oriented Programming, R2016a, MathWorks.

Index